Praise for Peter C. Stone's Illustrations:

"Sumptuous . . . beyond the tactile into a nearly mystical realm . . . "

—Publishers Weekly

"Dreamy . . . evocative . . . "

—School Library Journal

"Artist Peter Stone captures our essential connection to place in his powerful landscapes and words."

—Cape Healing Arts

Praise for *The Untouchable Tree*:

"Brilliant! Beautiful! A powerful vision that will nurture our abilities to think creatively and dream a peaceful sustainable world into being."

—John Perkins, New York Times bestselling author of *Confessions of an Economic Hit Man* and *The Secret History of the American Empire*

"*The Untouchable Tree* is a journey into the living energies of our world, whose patterns and subtle rhythms awaken cosmic and earthly wisdom. Peter Stone sees from the heart, through the eyes of a shaman. He invites us to do the same."

—Llyn Roberts, M.A., author of The Good Remembering; co-author of *Shamanic Reiki*

"Provocative and mesmerizing! Peter C. Stone's lushly layered paintings transport you into a dazzling dream world."

—Anita Winstanley Roark, director, Winstanley-Roark Fine Arts

"A body of paintings that employ a self-taught technique the artist refers to as Tonal-Realism, which incorporates the aesthetic of plein air painting with a complex and sophisticated layering of glazes. The result brings a visual and physical texture to the surfaces, while careful placement of varieties of saturated colors create a visual vibrancy and luminosity that is nearly palpable."

—David B. Boyce, *The Standard-Times*

"His brilliant landscapes contain elusive treasures as well as stunning vistas . . . Stone uses poetic prose and multi-layered paintings to evoke an understanding of the sanctity of the indigenous cultures' relationship to the earth."

—Terri Lerman, *Cape Cod Life*

Praise for *Sanctuaries*:

"An extraordinary, spellbinding book . . ."

—Gay Matthaei, author of *The Ledgerbook of Thomas Blue Eagle* and winner of the 1995 Christopher Award

"His colors dance and shimmer; the images entice the eye and invite exploration. . . . The longer one studies these paintings, the more one sees, and the clearer one's mind feels."

—*Fearless Reviews*

"Mr. Stone's ability to match evocative words with evocative scenery is haunting . . ."

—*Heartland Reviews*

WALTZES WITH GIANTS

The Twilight Journey of the North Atlantic Right Whale

Peter C. Stone

Skyhorse Publishing
New York

To Miss Cornell, For your Journey!

~ Peter C. Stone

In memory of my father
All the better for seeing you . . .

Skyhorse Publishing books may be purchased in bulk at special discounts for sales promotion, corporate gifts, fund-raising, or educational purposes. Special editions can also be created to specifications. For details, contact the Special Sales Department, Skyhorse Publishing, 307 West 36th Street, 11th Floor, New York, NY 10018 or info@skyhorsepublishing.com.

Skyhorse® and Skyhorse Publishing® are registered trademarks of Skyhorse Publishing, Inc.®, a Delaware corporation.

Visit our website at www.skyhorsepublishing.com.

10 9 8 7 6 5 4 3 2 1

Library of Congress Cataloging-in-Publication Data is available on file.

ISBN: 978-1-62087-106-5

Printed in China

CONTENTS

AUTHOR'S NOTES

One premise guiding this story is that the arts play a crucial role in helping develop our intuition and observational skills in the sciences. Both convey information—art transmitting feeling while science explains it, as E. O. Wilson notes in *Consilience* (1998): "Neither science nor the arts can be complete without combining their separate strengths." But the STEM (Science, Technology, Engineering, Math) education model falls dramatically short without Reading and Art to nurture our various intelligences and our humanity—our compassionate abilities to see and feel, infer and express. We must actively support the movement back to a STREAM (Science, Technology, Reading, Engineering, Art, Math) of disciplines.

The common name "right whale" was historically applied to the "black whales" of the genus *Eubalaena* and to the bowhead (*Balaena mysticetus*), also known as the Greenland or Arctic whale. Because of uncertainty in the nomenclature due to differences between present-day and seventeenth- or eighteenth-century perspectives and language usage, the name "right whale" here refers only to *Eubalaena*.

The North Atlantic right whale (*Eubalaena glacialis*) called Arpeggio, chosen partly because of her name's musical connotations, is inspired by a real female whale born in 1997 and chronicled in the New England Aquarium North Atlantic Right Whale Catalog. Her age, sensations, encounters, and the migration or "longwater" around which the story unfolds, are imagined. In the mythological view, an individual's name represents a lifetime of experience, along with that being's inherited wisdom. Arpeggio is considered here to be a creature that carries the same name or knowledge—that genetic information or evolutionary wisdom—as her ancestors.

The idea of mysticism is not only about inspiring a sense of mystery and wonder, but concerned with subjective experience of the realities not apparent to the two-legged sensory world. The word "divine" is used to mean "supremely good or beautiful" or "magnificent"; "mysterious" connotes that which arouses wonder while eluding explanation, in supporting the fabric of the mystical creature world (*The American Heritage Dictionary of the English Language*, Third Edition. Boston and New York: Houghton Mifflin, 1996, 1992).

The glossary includes both cetacean terms and two-legged words. The cetacean terms are inventions or translations for sounds that might describe energies, actions, creatures, events, names, places, or states of being. Many of these words are not found in books for readers twelve years of age and older. They are used here with deliberation and reverence. For to learn the music and mystery of new and different words, we must read them and hear them spoken. By the same reasoning, to realize and respect the languages of other cultures and other species, we must also learn to see and listen to the mystery and music of *their* words. Finally, to appreciate and respect the languages of environments that sustain life on this planet, we must learn to see and listen—more than ever before—with a sense of natural humility.

Mysticism and science meet in dreams.

—Edward O. Wilson

We need another and a wiser and perhaps a more mystical concept of animals . . . For the animal shall not be measured by man. In a world older and more complete than ours they move finished and complete, gifted with extensions of the senses we have lost or never attained, living by voices we shall never hear.

—Henry Beston

To get the feeling of what it is like to be a creature of the sea requires the active exercise of the imagination and the temporary abandonment of many human concepts and human yardsticks. For example, time measured by the clock or the calendar means nothing if you are a shore bird or a fish, but the succession of light and darkness and the ebb and flow of the tides mean the difference between the time to eat and the time to fast, between the time an enemy can find you easily and the time you are relatively safe.

—Rachel Carson

Every animal knows far more than you do.

—Nez Percé proverb

FOREWORD

They give us life.

We depend on their health for our own, for the very air we breathe. They cover more than 70 percent of the Earth and contain nearly all her water. Yet, while we have walked upon the moon and developed technologies capable of turning the globe into a wasteland, we have explored less than 5 percent of the oceans and named only a fraction of the species that live there.

Why should we care about those plants and animals most of us will never see? After all, a long parade of them has disappeared over the course of history. What's the worry?

Some contend that *Homo sapiens* is part of the system, that human activity by which other plants and animals perish is simply an evolutionary force. Even without the impact of *Homo sapiens*, some are eager to note that numerous species have already gone extinct. But these arguments are flawed, because survival of the fittest applies to Darwinian rules of natural selection on a cosmic scale. And in the 3.8 billion years of life on this planet, we are now in the throes of the first mass extinction actually *caused* by a living thing—the creature we call *man*.

Over the past fifty thousand years, as Paleolithic peoples evolved into the deadliest invasive species armed with technologies, agriculture, and disease, rates of extinctions skyrocketed. By the 1990s, assaulted by the demands of our industrial civilization, an estimated 27,000 species were dying off each year. Twenty percent of Earth's species are predicted to be lost by 2030. Near the end of the millennium it was calculated that more than 137 species (of plants, animals, and insects) were vanishing every day in the Amazon rainforest alone—the equivalent of fifty thousand species a year! And those that have been dramatically reduced in numbers, even if not at risk of being extinguished, have become genetically handicapped, losing their previous resilience and adaptability. In fact, unless humankind makes significant changes in the way it treats the planet, *half of all species* may be extinct by the century's end—the evolutionary blink of an eye.

So what do we do? Can we afford to worry about rare birds or bugs or bears when we have so many more immediate problems to be concerned for? Considering the undeniable interconnectedness of life forms throughout the six kingdoms, we do not have a choice. "When we destroy ecosystems and extinguish species," says the eloquent biologist E. O. Wilson, "we degrade the greatest heritage this planet has to offer and thereby threaten our own existence."

It is true that we also struggle with complex politics, instabilities, and threats that our culture addresses through our sense of National Security. But no less important are the biodiversity, ecosystems, and climates that sustain us. If that's the case, why do the goals of scientists and environmentalists often appear to conflict with those of industries and governments? And why does the work of scientists and environmentalists sometimes appear to support the unsustainable practices of industries and the governments that represent corporate interests?

Is part of the problem how we *define* National Security?

Before we regard a species as less essential than our systems of national defense, should we look at the consequence of the loss of that species upon its food chain and ecosystem? Should we question, in turn, what effects that altered ecosystem has upon the health of the oceans that impact this nation, the fisheries, industries, and associated economies that our culture values and wishes to defend? And dare we ask what pressures our consumptive society places on the oceans that keep us alive? Could the answers change our perceptions of National Security?

Many commercial fisheries and their management practices have demonstrated little concern for the collapse and loss of species, traditionally referring to fish as "stocks" and viewing them as harvestable natural resources. Catch share programs, proven successful in some fisheries, have not been widely accepted. Yet, since the 1950s, ninety percent of the large ocean fish have disappeared. Businesses have clearly not been required to bear their true costs when research has shown that by 2010 over three hundred thousand dolphins, porpoises, and whales were dying as by-catch or from fishing gear *each year.*

Why should we care about mammals entangled in lobsterpot lines and not about the lobsterman trying to make a living and feed his family? And if some whales abide lower down the food chain, how important can they be? What could be more imperative than the U.S. Navy's ability to train and test personnel and weapons, ships and submarines, without being restricted by the presence of something that few of us will ever see? Is it possible to even *know* whether an animal is vital to the health of the lands and waters that nourish us?

Can we afford to take chances and risk that animal's survival?

"The trouble," writes farmer and author Wendell Berry, "is that we are terrifyingly ignorant. . . . The acquisition of knowledge always involves the revelation of ignorance." Only recently have we begun to unravel the engineering marvels of mussel adhesives and spider threads that possess strength and flexibility beyond anything man can synthesize. The burgeoning field of biomimetics looks to the wisdom of living systems for solutions, from the energy efficiency of purple bacteria to the vast medicinal storehouse of rainforests that have been appreciated by indigenous cultures for millennia. If anything, these relatively new advances in Western knowledge should raise questions, such as: Is it our right to vanquish any other being from existence? What are the costs of throwing away several hundred million years of evolutionary wisdom that each species possesses?

"It does not matter whether it can be scientifically proved that life as we know it is in danger," explains Cayuga Bear Clan Mother Carol Jacobs. "If the possibility exists, we must live every day as if it were true—for we cannot afford, any of us, to ignore that possibility."

In other words, can we afford to miss the real-time lessons learned from an organism we have studied in great depth? As author Paul Hawken emphasizes, "We remain largely ignorant of how the infinitely complex connections between different biotic communities affect the well-being of all species, including human beings."

Perhaps we need to regain a sense of wonder for the world around us, to observe and imagine, and to learn some humility and respect for the things we do not know.

That is the reason for this book.

Waltzes with Giants is inspired by a North Atlantic right whale named Arpeggio and the many people who have worked diligently to protect her threatened migration routes along the eastern United States as well as her calving grounds off the coasts of Georgia and Florida. The aim is to evoke the wonder, the sorrow, and the conflicts associated with this member of the suborder Mysticetes, the baleen or "whalebone" whales. For we have long been a part of its story.

Slow and harmless as a mouse (when left alone), the North Atlantic right whale (*Eubalaena glacialis*), like its counterparts the Southern right whale (*E. australis*) and the Pacific right whale (*E. japonica*), was found by whalers to be rich with oil and baleen, making it commercially sought-after. Once killed, its buoyant blubber (like that of humpbacks or the more valuable bowheads) usually kept it afloat for the taking. It likely earned its name because baleen, the fringed plates of horny whalebone hanging from the upper jaw, was seen as an indicator of a "true" or "right" whale to its hunters—until it became a symbol of too much hunting.

Hunting is part of our complex relationship with the oceans, one that has grown fragile since the Industrial Revolution began to magnify our excesses for extraction and pollution. We ship and consume and hunt with abandon, and the oceans suffer as a result. The whaling industry and other fisheries have historically decimated the populations of fish and whales upon which they depend. Where many thousands of North Atlantic right whales inhabited the seas before the height of the whaling era, now only a few hundred remain.

But other factors have also contributed to their decline, including loss of habit and competition from bowheads. Whatever the possibilities, scientists and conservation groups consider the right whale to be a "canary in the coalmine," a species that helps indicate the relative health of the oceans upon which we depend for the air we breathe and the food we eat. This idea can be applied to any species and unstable ecosystem, and ecologists say that unstable ecosystems are ripe for change.

With that in mind, years of conflicts started coming to a head in the new millennium between conservationists and commercial interests; scientists and politicians; and the courts and military initiatives. Scientists were blamed for not having enough convincing data about the North Atlantic right whale. Lobstering and fishing businesses demanded their right to operate. Environmentalists often appeared to be working only reactively to undo the impacts of human activities. The U.S. Navy deemed realistic active-sonar testing crucial for anti-submarine warfare training, while there was little conclusive evidence of sonar affecting baleen whales the way it does toothed whales (of the suborder Odontocetes). And commercial shipping interests bemoaned restrictions on ship speeds to prevent collisions with right whales, when evidence of the negative impacts of ship noise was mounting. All of these groups (artists and writers as well) play roles in servicing our industrial civilization that is relentlessly plundering the biosphere.

In the balance swims a gravely endangered mammal—a nuisance to some, an obstacle to financial gain for others—as its sonorous voice is being drowned out by our quarrels.

Certainly, the clash of man versus nature is nothing new. It has been replayed constantly for centuries. That is why this book is not intended as an ominous biography of yet another vanishing species. For indeed, which is more dispiriting: telling about a creature on the brink of extinction or relating that so many species are being wiped out by human activity that the phrase "on the brink" has become a colloquialism?

Instead, *Waltzes with Giants* strives to pose questions. When we ask, "What if?" can the answers help us treasure the importance of "What is?" Why not embrace art and science together to become engaged in the mysteries of the oceans? If we value them, will we save them? Are we capable of finding solutions that protect human interests by protecting fauna and environments integral to those interests? Can we demand that industries take a more responsible position by internalizing their *true* costs of doing business? And rather than insist that our sciences and technologies keep trying to solve the environmental problems we have caused, why not employ them in our creation of a new cultural relationship with the environment? "New technologies and new habits offer some promise," suggests environmentalist Bill McKibben, "but only if we move quickly and decisively—and with a maturity we've rarely shown as a society or a species."

In essence, can we actively dream a new dream of the way the world should be? To help answer these challenges and transform ourselves, we will need our compassion as much as our sciences. As the visionary Paolo Lugari states, "There's no such thing as sustainable technology or economic development without sustainable *human* development to match."

By listening to our humanity we may discover a new relationship with creatures like the right whale. In order to do so, we must move beyond faith-based environmental activism to a covenant built on the unity of a living planet. It will be an association fueled not by hope and its complacency, but by desire—the *urgent* desire for humankind to understand our critical place among the miraculous fields of nature.

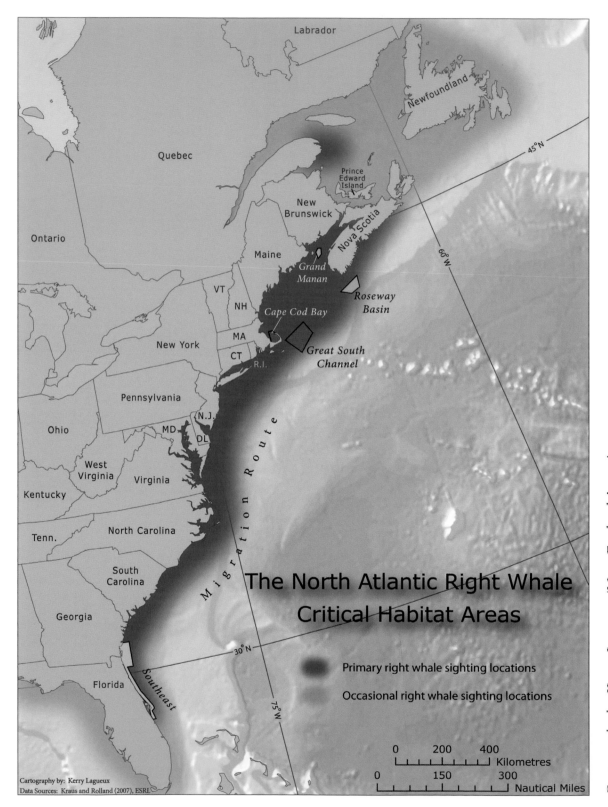

The North Atlantic Right Whale
Critical Habitat Areas

Primary right whale sighting locations

Occasional right whale sighting locations

WALTZES WITH GIANTS

Chapter One
Imagine

Hush.

Be still.

Listen!

A muted lowing . . .

A looming presence . . .

The legato purl of rising bubbles . . .

The feathery sweep of sandpaper flesh through cold blue water . . .

Shhhhhh . . . Listen!

IMAGINE   3

Why?

You might heave a deep breath of damp salt air, or blink at the sunlight glancing off towering ice castles.

You might be amazed by all you can hear—before you glide beneath the waves.

If you would only imagine *why* . . .

If you could, would you ask?

Why? What is the use of imagining?

Try it.

Imagine you are the only mother you know who is nursing a child.

The only sister left who still has a brother with whom she can sing.

The only daughter swimming who remembers the colossal form of her father.

The one upon whom all the children of your children depend.

How can a thought so bleak come to pass?

Imagine.

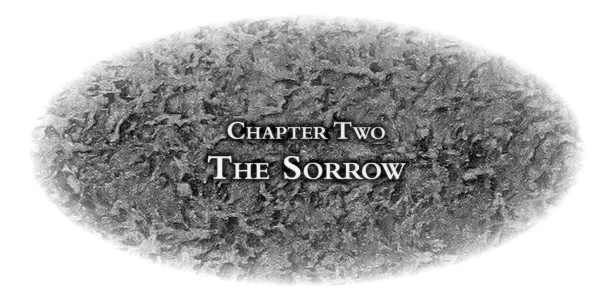

CHAPTER TWO
THE SORROW

During the longwater, a silver light flutters in your mind like the staccato flash of a minnow in a tide pool.

It doesn't surprise you though. The annual passage always reveals stories about the past that tell of legends from the future.

What has been the trigger this autumn?

The sight of your immense brothers and sisters swimming off on their solitary migrations? Their tonal rallying cries?

Your youngest daughter drifting nearby?

Or have you been prompted by something simpler, the swirls of plankton in the blackwater? The bark of a young seal or the *CRACKKK!* of melting icebergs? A pink flurry overhead with the whir of schooling krill? Or the weakening warmth of the Mother Light and the lengthening interludes when she sleeps?

What awakens your memory this season?

Is it the familiar forms beneath the waves, the mountains that climb to become islands where the currents run foul? Do you recognize the acid taste of rivers disgorging their nitric effluents, the clouded thinwater where nothing wants to swim?

Or does the sheer emptiness of the waterfields strike you, the realm where once thrived more creatures than your dreams could conjure?

Have you ever wondered whether nothingness can become as powerful as sorrow?

What is this unspoken stirring in the awareness of all your kind?

Does it come in gloomy choruses from the breedwater that should be blessing every womb? Or the depleted genetic ability of your pod to adapt and persevere?

With only a few mothers carrying their unborn more than halfway through the twelve to fourteen months of gestation, there is no telling what the future holds.

Whose silent voice spins your inner compass to guide you south again toward the birthwater? Are you distracted by a blinking beacon on the bold cliffs of what the two-leggeds have entitled the *New-Found-Land* ?

How perplexing, their conceited names, as if they actually believe that they were the first to discover the shores along which you have swum since memories began.

Not many of your kind come this far anymore, to where the ice islands linger. Most gather in the blackwater known to the two-leggeds as the Bay of Fundy.

But this year you lost yourself, groping blindly through soiled currents of two-legged shadows and noise. If you swam astray again, could one of their "houses of light" ever guide you on your way?

No. The summons is more compelling than that. Just *listen* to the sound of it!

You feel again an emptiness so deep that you let loose a long contralto wail—a requiem for your kin who have been run down by the shadows above or strangled by the silent black serpents that drape down through the water column.

So many times you have tempted fate and brushed by them. Steely and beguiling, they snake up from baited traps on the bottom like naked stalks of giant kelp.

You know the shadows have put them there, to snatch away the eight-legged crustaceans and others— the ones the two-leggeds have named the *conch* and *hagfish* and *black sea bass*—never to be returned.

You are simply aware of an unsettling peace as you meander through jungles of menacing nooses and spidery webs that lace the thinwater. Looping between the two-leggeds' trawls, the wire mesh traps strung together along the bottom, they sway in the surge to snag you,

bleed you,

drag you,

choke you.

You may be strong enough to swim off, tangled in the gear, but it can starve you into the cruelest silence. Even if it takes the full turn of a longwater.

Twice the lethal groundlines have wrapped and gashed the flesh of your flukes. Still, you swim undaunted, for any premonitions aren't as important as the beauty with which you move.

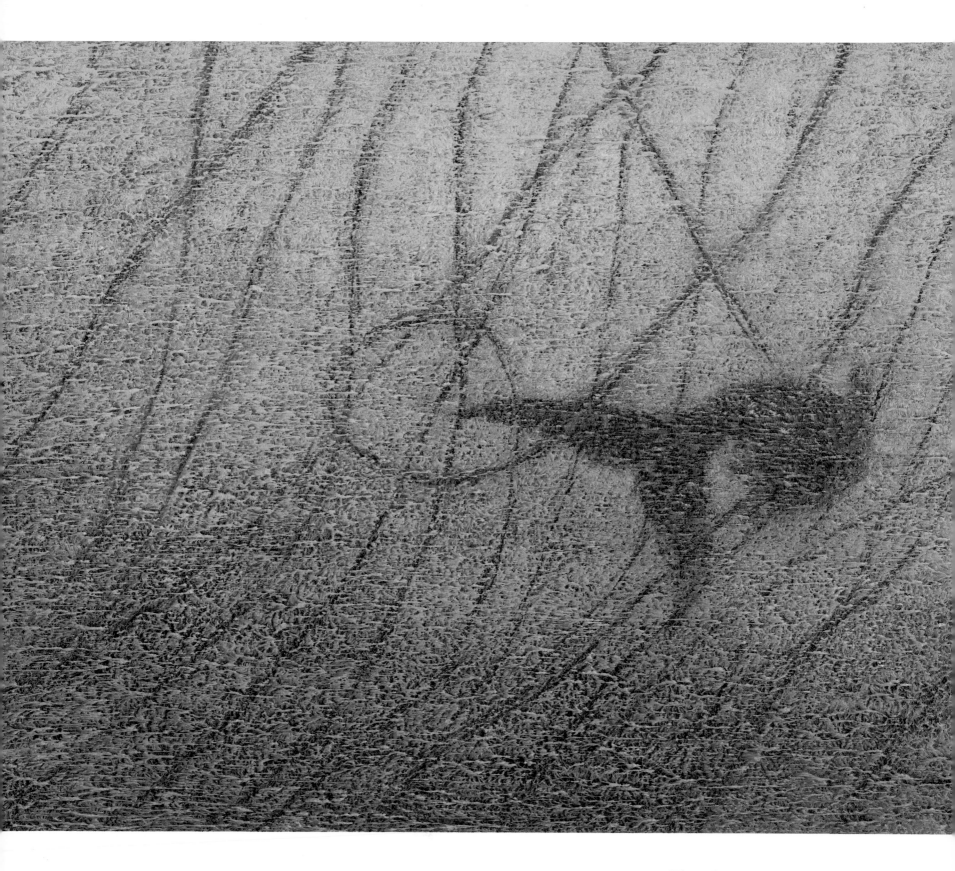

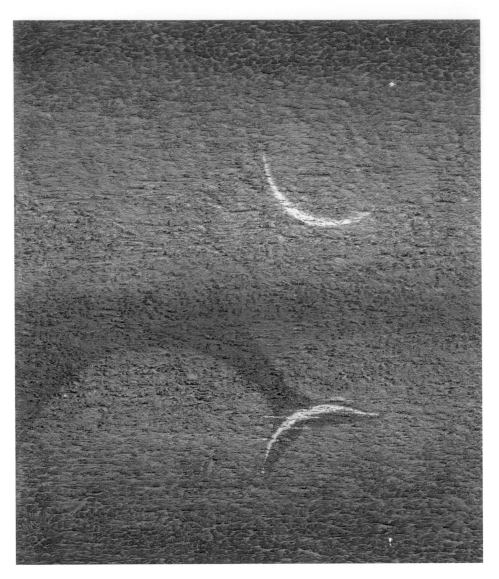

You skim the swells and dive, feeling the thrill as your bristled baleen strains the copepod water. The scattering creatures on the surface give the appearance of a downpour. Waves sluice through your gaping maw, spilling up and over your mammoth arched tongue. They gush back out by your eyes and strand the high energy-hordes for you to gulp down, a million in a mouthful. Soon you are nodding through the swells to clean off your baleen after your meal.

Is it possible for each day to begin with that kind of joy, no matter how dark the night before? When your progress is as pure as that of a crescent across a tranquil watersky?

Who knows how many two-legged wastes you ingest at the same time? They do name them with strange sounds—*plastics* or *pesticides*—like sticky stuff caught in their teeth. The thought doesn't occur to you though. Those are two-legged ideas.

Homing in on the densest concentrations of prey, you swoop and wheel in the slow-motion arcs of a giant albatross. All as you warily keep in mind the shadows above, the *shadows that move with thunder*.

Experience has taught you to identify their dissonance. It feels dry and jagged to your sensitive ears. Though it may be a harmless sound in the two-legged world—they know not the pain you felt on the day they took your brother

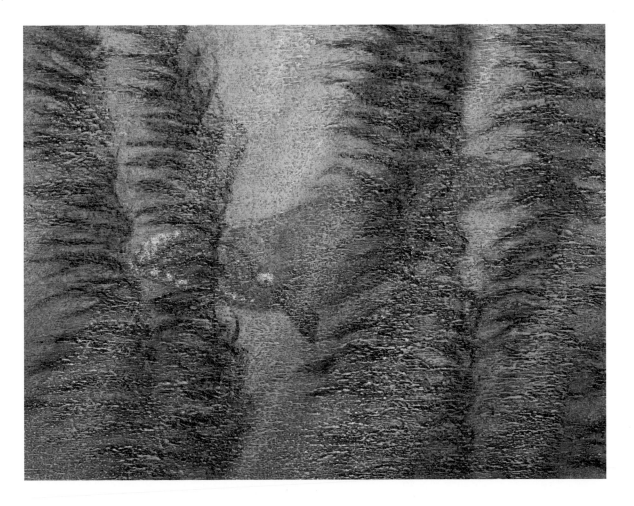

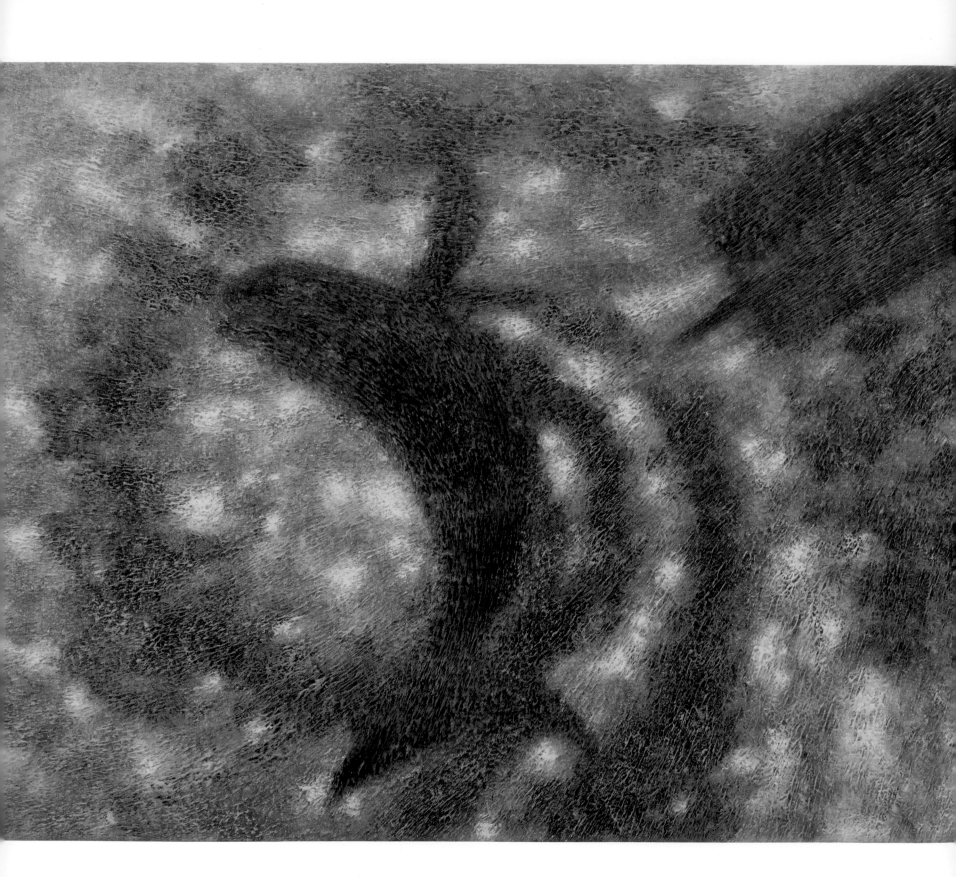

14   Waltzes with Giants

You cannot forget his panicked trumpeting, *NMmooouuu!*

It was deeper than a foghorn, an elephant's urgent basso profundo with the sorrowful lament of a loon. It scarcely resembled the sacred sound that all whales sing, a melody woven with the tides and the Dream Light and the womb of the Mother.

Your brother's final bellow came when a droning phantom charged into him. The monster carved his side wide open, its razor teeth gnashing like bluefish in a feeding frenzy.

But it wouldn't stop. It kept attacking! Slicing! On and on! Relentless as a buzz-saw! *Slicing* in a whirl of sabers!

You still hear his angry cry that he could not flee or protect the rest—just before the looming keel broke his back in two. You weep as you picture him hanging there, dangling upside down after the ship strike. You still listen helplessly to his diminuendo groans that he would miss you, spilling red clouds and writhing in spirals.

Yet he was unable to sink where he wanted to go, down into the peace of the blackwater. Because his own buoyancy lifted him in agony to the watersky, and on to the Between.

Do you think such a ghastly execution could ever happen again?

Chapter Three
Remembrance

*C*ome back! cries your silent voice. *Nmmooouuu! Come back, my brother!*

The sorrow stirs your memory to unfold like a swaying anemone, in the way that all cetaceans experience the dreamworld. You see yourself within the first flicker of your soul, the spark at the beginning of being for any sentient creature.

In dream consciousness, you travel on both sides of the watersky. You shimmer and glow like a candle brought to life, burning firmly yet unrippled by any wind. In fact, you perceive everything touched by the web of water, for this is how you remember.

You are aware of the liquid warmth of your mother's womb, her careful movement even when hungry and wounded. The vibrations of her voice soothe you, the embrace of her body. She feels so safe and tender, despite her seventy tons of streamlined girth and blue-black fusiform longer than ten two-leggeds laid head to foot.

It is the womb of all your grandmothers you recall, encoded by genes passed on to you by your ancestors— those who returned to life in the waterfields while your hippopotamus cousins remained in the dry realm.

And now it is your only sanctuary.

You slide past what the two-leggeds call the rocky Scotian shores, scoured many longwaters ago by receding glaciers that challenged your grandmothers during and after the Long Cold. Did they suffer then as their habitat changed? Was it the beginning of their decline? For their domain once included every northern edge of the bigwater.

On you cruise through basins that two-leggeds have named Roseway and Grand Manan, the blue depths of the grandest tides that rise and fall in heights almost equal to your length.

Most of you gather here before the longwater, sometimes cavorting on the surface in active groups, nuzzling prospective mates, roughhousing through the waves.

Into the Gulf of Maine, you and your grown daughter slip silently under island reflections of spruce and granite. Does it ever cross your mind how few two-leggeds see what is going on beneath? You sashay over sea oak and red fern and horsetail kelp as your grandmothers have always done. You know nothing of those two-legged descriptions—you only recognize that you have come this way before, descending by green fleece and sea colander and chenille weed.

On deeper dives you might even scrape the bottom while searching for food, emerging later through the watersky with your snout and bonnet smeared in thick brown mud.

You joyfully dodge the treacherous black serpents, ducking under stray tangles of torn gillnets and floating groundline, and the merciless longlines spiked with barbed hooks. You dance past bulbous fiberglass buoys that trail knotted snarls of polypropylene rope like the stinging tentacles of some prehistoric Portuguese man-of-war.

Being in this place again evokes the warm milk-taste of your mother who nursed you during your first year, letting you suckle beneath her smooth broad flippers.

Far from the rugged coast you breeze through Jordan Basin, another label that has no meaning to your kind.

But the currents do. They carry secrets of magical winter congregations when your listless brothers don't appear to be feeding or diving; not even resting up for the longwater and bobbing side by side like waterlogged timbers.

Are they simply waiting? Or is this really an ancient mating ground, a place for your sisters' calls to be heard?

You swerve like a hydrofoil and ram into an eddy of foodwater. Your steam shovel-mouth yawns wide to scoop, letting water pour over your furrowed tongue. The succulent waves boil and strain through your fringed baleen, the horny whalebone that hangs in perfect filtering curtains from your snout-like upper jaw—your rostrum, as the two-legged wise ones call it, in their glee for scholarly words.

Would it amuse you to know that your baleen reminds two-leggeds of a rather large, floppy mustache? Or aren't whales so easily amused?

Don't you relish the tickling sensation from the juicy swarms of captured zooplankton that will nourish you, the tiniest creatures a carnivore could hope to feed upon?

They don't have a chance.

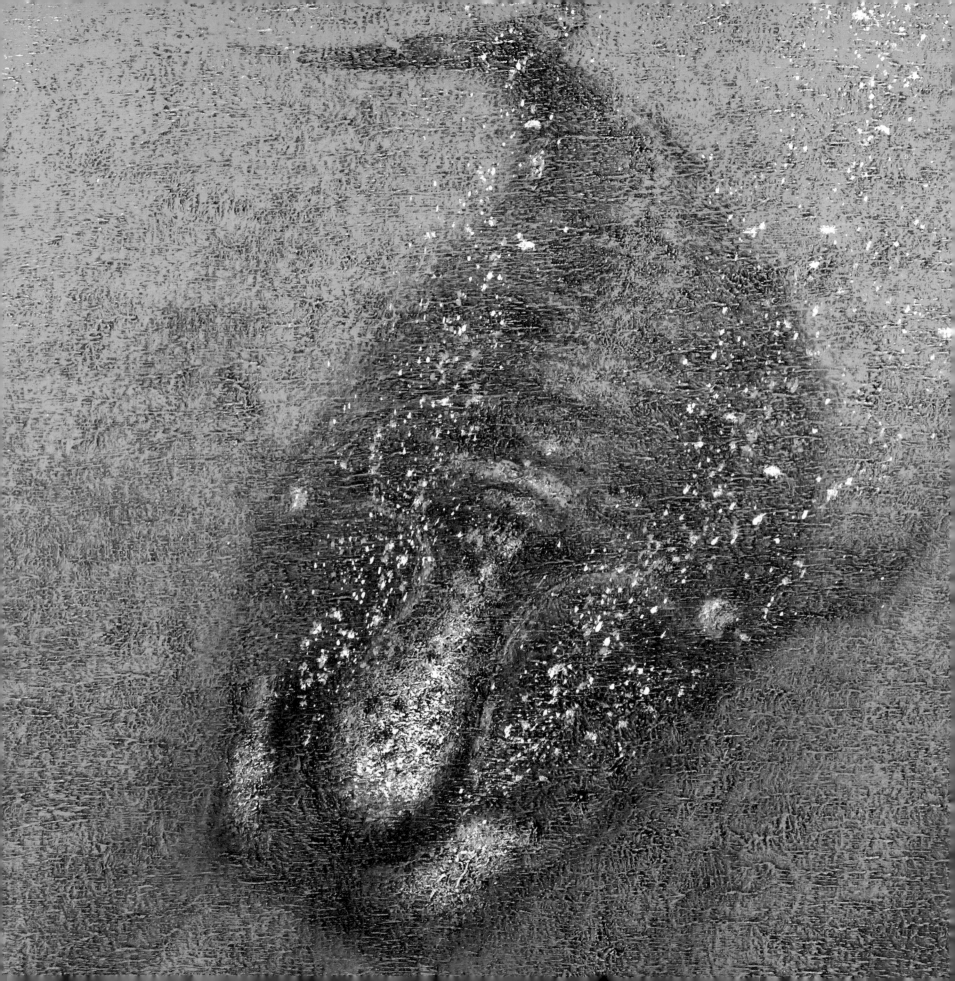

22

Will you ever forget the tales your grandmothers told of the huge shadows that moved without thunder, without warning? How they twisted their heads at the puzzling sight of those ominous specters stealing across the glittering watersky?

The enormous aliens' dark bellies were scabbed with barnacles and green scum. They came from the eastern shores of the bigwater where they had been slaying your kind with ease. Their stubby vertical flukes did not sweep up and down through the water in the manner of whales. Instead, they barely shifted, ruddering from side to side like the caudal fins of primeval hunters. Some splashed their many pectoral flippers together in a peculiar cadence. Others had tall dorsal fins that rose into the above like icy curtains, flapping and filling with wind.

They flung kiltusks sharper than the unicorn horns of narwhals that speared your grandmothers' flesh. These trailed serpents as long as tidelines, tied with wooden barrels to keep your grandmothers trapped on the surface of the watersky, spouting ruby mist, flailing and sobbing until all their strength was gone. Encircled by furious gray fins and raging jaws.

Before they were hauled back and strung up, struggling, bleeding, drowning with shock.

Into the waiting death camps where horror poisoned the air.

There was no way to escape the aliens' butchery. Their boarding knives and spades and flensing blades flashed and fell until the tides ran red. Their blubber-stoked fires crackled, spitting sparks and billowing grease-smoke.

There was no hope of surviving the flaming tryworks, the boiling cauldrons of your grandmothers' life-giving oils that would fuel the gluttony of the civilized two-legged nations.

Your memory sees what your grandmothers then realized, that the great shadows were trained and ridden by the creature you had begun to fear more than all else. And the same buoyancy by which you could rise swiftly to the watersky had become a mortal trap. It slowed your grandmothers' ability to escape into the depths. It left them exhausted. Defenseless.

Could you believe your grandmothers' stories? The ice-skinned two-leggeds on the shadows from across the bigwater were unlike any other living thing. They killed not just to feed themselves or their families, but for profit . . . and for the glory of dominance. Wielding irons, pikes and grapnel, poisoned arrows, fatal lances, they appeared bent on vanquishing your kind from the waterfields.

Their hunters shouted unintelligible sounds. "*Black whale!*" they called out when they spotted you.

"*Fire-in-the-hole*!" they cheered, after plunging barbed kiltusks into your grandmothers' hearts.

"*Chimney's-afire!*" they exulted, thrilled by geysers of bloodied vapor from the blowholes of the wounded.

But it made no sense! They acted as though this was a form of victory. Did it seem worse than ignoring the wisdom of your elders, or forgetting that remembrance could be a teacher?

Eventually, your grandmothers assumed that the huge silent shadows died off themselves. Who knows where they vanished? Some may have been frozen into the ice for many longwaters before the above began to warm, and the sea ice began to melt, and the waterfields began to rise and consume the dry realm.

That was the supposed way of the balance, according to the power that moves in circles.

Until another large predator soon made its tracks across the bigwater.

Was it hard to fathom that a shadow could be so much more ruthless?

Your grandmothers recognized it as kin to the first wicked intruders, but it had no flapping dorsal, no stubby pectoral fins. The flukes of this nightmare churned the waterfields and belched roiling black fog that cast an acrid pall over the watersky and clouded the Mother Light.

You could smell it long after you dove for the depths. It was big and fast enough to chase down the swiftest of cetaceans, the baleen families of fin and gray and blue; the toothed families of beluga and sperm and

narwhal. And its kiltusks foretold the future's wrath, topped with swivel-head grenades and unleashed by the two-leggeds' cannons.

After that came still more massive savages, the knife-edged shadows and their whining metal flukes. They stampeded across the watersky, leaving sooty clouds in their wakes.

These shadows moved impossibly against the rhythm of the waves. Their grinding noises were so staggering that you came to know them as the *shadows that move with thunder.*

They behaved as though their formidable show of brute force was proof of wisdom, bolstered by the cutting racket of iron on iron, steel cables screeching around winches, banging drums and clanking gears, the thrumming engines of war.

You recall the fear in your mother's violoncello voice as she described the terror of the madsounds, the panic in her heart, the blood uncoiling like crimson smoke from her sisters' wounds. Did she possess the ability to question?

> *Why didn't your brothers learn about this? Haven't you warned them to look up? Look up! Look above you to the watersky! Don't you see? LOOK UP!*

Did you sing back your protest, repeating some sounds in couplets as right whales do?

> *Look Up? Of course, I've told them! LOOK UP!*

Perhaps you have communicated to others that the warmongering thundershadows and their death knells come from the two-leggeds, although you cannot explain how or why.

You just know the madness is true.

WALTZES WITH GIANTS

A few of your kind don't seem to accept this. And some of your brothers act oblivious, more intent on showing off their presence by slapping the surface with flippers or flukes and popping off a string of gunshot sounds.

The madness is true? Is it true?

Might they answer that way, if it meant something to them?

Still, others drift away without acknowledging, because the two-legged way of reckless destruction has no logic in your world. The bewildered voices of so many whales often convey their disbelief. If their voices were to become two-legged words, what would they say?

You wish, Arpeggio. Think what you wish, but no creature would ever wantonly vanquish another . . .

Why? Why, could that break the circle of the seasons?

How absurd. Absurd! It has never happened before in all of Being!

Wouldn't that defy the wisdom, the wisdom of the grandmothers?

You have pleaded with them, unable to clarify.

Look up! Above you! Please, LOOK UP!

Now, it is too late.

So many of your friends are gone, and there is no forgetting that gruesome night.

CHAPTER FOUR
THE UNDOING

After the Mother Light had vanished, it was the *madsounds* that drove your pod insane.

Some of you were already wrapped in sharp netting, trailing frayed lines. But nothing could prepare you for the onslaught, the rising wave of noise from the *shadows that move with thunder*.

A devilish *pinging* swept down through the water column. Pounding in pulses, frying your senses. A terrified trio of pilot whales sped past, whistling and clicking nonsense, their eardrums punctured, their tissues ravaged by bubbles of gas in the sickness two-leggeds call "the bends."

You had no knowledge that toothed whales might be affected differently than your kind. Only that you grew dizzy with a need to escape. The lines cutting into your flesh made you panic. Icy knives stripped your nerves. Your burning lungs shriveled. The stress froze your bones. Your splintered mind spun you around and around.

Can any song express what you witnessed? Dolphins in dark currents twisting and lurching, undone by the pain. A child floating slowly to the surface without answering its mother's call.

Then you lost track of your sisters, as their cries became smothered by *noise*. Yet beneath the murky glow of the Dream Light there was nowhere to go, only away . . . away . . .

A flurry of beaked whales fled there too, followed by the stricken dolphins—*away*.

Until the thinwater couldn't hold them anymore.

Until the shallows scraped their bellies and they launched themselves up through the watersky, skidding across seaweed and sand.

Before long their organs began to hemorrhage beneath their own weight, and their lungs started to collapse.

Where was this forsaken "*away*"?

Already their dark flesh was absorbing too much of the Mother Light's radiance.

Their flukes arched stiffly into the above. Their core temperatures spiked out of control, boiling them from the inside out.

And all you could see from a distance were reflections of fire and blood.

True, the madsounds had stopped. But it was a terrible silence.

Thankfully, you did not follow. How could you? Flailed onward by the snares of netting, each razor strand, each strangling groundline?

Was it pure panic that made you lose track of time?

Perhaps that is why you didn't notice the stealthy approach of a smallshadow. A tugging at your back. A whining buzz of smallshadow noise. A thin silhouette held high that resembled a kiltusk. A long silver stinger the size of a needlefish.

You startled at the jab, twisting and rolling to evade more discomfort.

You sped up and tried to push onward like a pilgrim on a quest, as a weird sleep trickled into your blood.

Gradually, though, your breaths came quicker and more shallow.

Then drowsiness crept over you like a warm blanket of sea jellies.

Before you knew it, the two-leggeds were right there at your side in the dripping twilight, leaning out from the smallshadow with long blades to undo the mess of suffocating gear.

These were the first two-leggeds you had ever encountered.

With a slow-motion blink in the grasp of their sedatives, you curiously eyed their finless bodies and puffy red torsos before recognizing something soothing in their presence. They weren't what your nightmares expected. Their voices told you that they meant no harm but you didn't understand their words.

They gently touched your rough dark flesh, silenced by the grim sight of deep white scars across your back. They cut away the fishing lines that wrapped and gouged one of your flippers. They brushed the yellow lice from your brittle callosities, the cornified markings unique to each of your kin.

The two-leggeds with red tops seemed to recognize the crescent-shaped blemishes like white birthmarks along your jaw and rostrum. They smiled; the caked blotches around your eyes resembled the face paint of a two-legged circus clown.

And they frowned at your festering sores caused by the sharp lines.

They studied the yellowish rake scratches radiating from both your blowholes, still riddled with the lice they call cyamids from your recent ill health and the entanglement. They acted familiar with the ridges on your lips; the bite marks on a flipper from an encounter with a rogue shark in your first season; the oblong shapes of pale coloration along your belly; the skin lesions on your flukes spreading outward like crusty green lichens.

One of them wrinkled her nose at the rotting fish smell, but she didn't turn away, apparently unafraid of your steam-engine-girth. She made a troll face at your foul sweet plankton breath, the gummy stench of sea lettuce and mudflats, your wheezing caverns burping up your last meals of oil-rich copepods.

You tried to speak but your voice was too weak and low-pitched to be heard. Besides, many of your pod don't believe you have the capability to communicate with two-leggeds.

Why should they? Your living relatives are too young and too few to know that the ancestors of *all* creatures once spoke to each other.

You are one of the elders now, fifty-eight longwaters with nearly twice that to come—if you can survive.

You are a yearling, an adult, a new mother, a grandmother. You are a wise one possessing the knowledge of your forebears in every muscle and bone, the details of your evolution—all carried in the musical name conferred on you by the redtops: Arpeggio.

And in your sacred name that never shall be spoken.

It is understandable why the younger ones have no way to envision the seasons' ebb and flow that you know as the circle. The currents of the waterfields no longer carry those lasting signs of equilibrium.

Gone are your grandfathers who had passed down the stories through sons and daughters, from calving to passing to calving in the endless round. They had recounted the times of constancy, a hundred thousand calvings ago, when the waterfields teemed with finned creatures of every shape and size.

From their memories you learned how the circle is as mighty as the Mother Light.

That is why, each time you rise to greet the watersky, you leave an ever-expanding ring as a token of your reverence—for purely being there.

CHAPTER FIVE
WATERFIELDS

What do you see when you breach? Do you laugh?

What do you hear when you crash through the air?

What do you feel when you shatter the watersky, splintering fog into soft bright prisms?

In the histories you tell with your timeless song, there were mud-skinned two-leggeds who lived along the shores you passed during the longwater's minuet. They must have known you were vulnerable in your shallow group swim. So they took your kind, as the orca or shark would sometimes take your calves, but usually just the easiest to catch: the sick and the lame. And only within the endless circle.

They brought death to what they needed for life.

The bigger and stronger of your grandfathers lived on to pass their seeds of the greatest wisdom down through the calvings. At the time, there were no thrumming shadows fast enough to run them down from one shore of the bigwater to the other. Nor did the two-leggeds' venomous wastes seep into the waterfields to kill things that grow and swim.

The balance of the circle came from its inherent truth that Being would continue. For the same circle of the watersky appeared as the Dream Light, and the tides, and the Mother Light. It was manifest in the shape of the great winds that blow, and the soaring wings of the air, and the ascending bubbles of all swimming things.

Before the ice-skinned two-leggeds flooded the dry realm and their numbers grew vast, their traditions seemed sustainable too . . . until some of them lost their way in the darkness of their own eyes.

Did you think they grew indifferent to the music of the waterfields or just lose the ability to listen? Maybe they forgot the delight to be found in a chorus of combers rolling across the watersky. What if they couldn't hear the thinwater melodies lapping against the shore, or your joyous duets you now sing and waltz with your latest daughter?

Did you see them neglect the power of the Dream Light shimmering above?

Any finned creature to peek through the watersky could behold the blooming greens and blues and golds. Is that why your grandfathers learned to rise, to savor those hues of fertility and wisdom and divine power that rolled forth on the waves?

The two-leggeds are not all bad, of course, you might admit—if whales admit anything at all. They do have their wise ones, the redtops, committed to saving your kind. But even they don't understand that much about what you actually *do*. Or *how* you do it.

They buzz after your kin on smallshadows and collect all manner of flotsam and jetsam. Even your waste, which has not once been wasted! They like to play tag, sneaking up to stick gadgets on your broad slick back. At times, they scoot by to dart your blubbery skin for samples, hoping for more clues from what they call the "code" of every cell. Would the chases would amuse you if they weren't so bothersome?

Others scratch their heads as if baffled by what you know in your bones, scribbling nonsense and gazing thoughtfully upwards with their four shiny eyes. Like pensive composers of a tragic libretto, they meticulously catalog most of your kind with pictures and numbers, tables and charts. All for an improbable opera.

At least the two-legged redtops have taken the time to understand that, once, there were more of you. You didn't just lurk alone through kelp forests, solitary as a shadow.

Clicking away with Cyclops eyes on shiny black boxes, they often speak as if you comprehend their primitive language and might one day reply.

They jot notes about markings and comments on scars, as if numbers and names from a two-legged contest—Snowball and Phoenix, Calvin and Starry Night, Piper and Shackleton—could ever replace the mysterious music with which you name each other.

Not that you object. They do mean well. They *are* trying to understand what they call *oceans*.

But don't you question if other two-leggeds could learn to ask, *When is an ocean not just an ocean?* Doesn't every creature understand that knowing words and names may be knowledge, but it is surely not wisdom?

Haven't they noticed? The birthwater has gone barren for many migrations! Few young have joined the circle! That is why your generation had no one with whom to frolic and dance.

You were the only calf that season. *The only one.*

And the hopes for this season rest on but a few solitary mothers making their way south.

Chapter Six
Man's Laughter

What do you glimpse when you rocket above, up through the center of every horizon, the circle where sky meets the sea?

Why do you breach with such ferocity?

Are you shaking off sea lice or just giving them a scare?

Have your flukes sculpted clouds and splashed over the Dream Light?

As you enter what the two-leggeds call the Great South Channel, you detect a salinity change at the edge of the fresh water plume flowing from the Gulf of Maine. You hear a smallshadow chasing a sister to release her from webs of gillnets and buoys.

You have seen red-topped two-leggeds use the floating barrels, too. Not in the way of two-legged hunters, but "kegging" to keep track of entangled whales who are crazed and confused, struggling and wasted. Until they are undone to the point of waterdeath, drowning to regain their freedom.

Soon you and your daughter pass the long sandy cape the two-leggeds have named for the finned one whose population they decimated. A fish once so plentiful that its schools could stop the shadows in their tracks.

Do you ever imagine what this place would look like from above? The two-leggeds even praise it as they would some grand deity, their almighty *Cod*.

Conceivably, the two-leggeds once praised your grandmothers as well. Until they slaughtered your kind by the thousands!

Why! you could protest—if cetaceans ever complained. *Why?*

Because you were so pathetically easy. Because you swam so shallow.

While feeding in groups you weren't hard to spot from a yardarm or crow's nest, especially with the distinctive v-shape of your blow. And after they stabbed you with kiltusks you bobbed back to the surface, wallowing belly-up like an overturned iceberg melting away.

How ironic, the oily blubber that kept you vital and warm now delivered you into the hands of your executioners.

They craved it voraciously.

That was why their shadow fleets skirted the sea ice, looking for your telltale spout.

At first they merely wanted you dead.

They coveted the wealth of your baleen, and your blubber and oil for fuel to heat and light their long dark winters. Baleen was the reason they proclaimed you a "true" whale to hunt, "a right proper whale," blind to the likelihood they were killing off your kind until there would be nothing left to satisfy what they proudly dubbed their "industries."

They desired your baleen for their vanities: things they called *corsets, riding crops, hairpins, umbrellas. Umbrellas?* What unusual sounds they rolled off their tongues. Would you believe they slew multitudes of your kind to help keep the rain off their heads?

And even after most of you had been forced from the waterfields, whether by hunting or loss of your habitat, the two-leggeds celebrated their cunning and know-how. Their societies flaunted it as a symbol of grandeur.

Would you have believed the romantic proclamations of their museums? "The Heyday of Whaling!" "The Era When Mankind Ruled the Seas!"

They were probably right. The waterfields bled into slaughterfields.

By that time, no greatfish had a shot at getting away.

But now they declare you're free to go? Now that they've killed you off until scarce hundreds remain.

Have two-leggeds given up the chance to witness your glorious beauty? Maybe they don't realize there is glory in all things, even those so needlessly beautiful.

If they did, Arpeggio, more may have sought to learn from the knowledge you carry, the secrets you guard. They may have opened their ears to your sonar symphonies, or to what you are saying when you call out in staccato bursts.

But they ignored the soothing bellows of your trombone voice.

They could have questioned your crackling gunshot sounds, emitted so frequently on dark winter nights, or your quizzical up-calls often whooping from low to high at dusk in the spring. Maybe they would be astonished by your rich harmonic moans as you dive a hundred meters, if they could dive so fast and deep as well.

So what if you have no vocal chords! That doesn't mean you can't sing!

Can you contemplate any other existence without the freedom to sing or dance?

If you gave it a thought would you ask,

Is that how the two-leggeds live? The two-leggeds live, without the freedom to sing or dance?

The adagio pace of the longwater takes you and your daughter along coasts with odd two-legged names; by the bay they know as Delaware, through thinwater where you recognize anemones and sea stars, sponges and horseshoe crabs.

But these same currents carry what two-leggeds call *litter* and *sewage, industrial chemicals, pharmaceuticals, fertilizers, fungicides, effluents from power plants, shipyards, factories*, and whatever else they wish to waste.

You plow through their carelessness and dive together over the Baltimore canyon, gazing down into the black abyss—realm of adventurous mystics and shamans.

Still, you journey on, near where the watersky burns sapphire into marshes of amber.

You swim past the great mouth of the Chesapeake and her suffering estuaries, laden with runoff from words you do not know: *restaurants* and *city streets, lawns* and *rooftops, farms* and *highways.*

How can you avoid the profusion of algal blooms and their potent biotoxins? Do the two-leggeds admit to upending the scales of these places, loaded with nitrogen, starving for oxygen?

Or do some of them speak with the tongues of eels?

Do you ever see things from their point of view, wading through tainted wetlands where herons step cautiously and the killing tides turn red and green?

It is doubtful.

For many of their kind are motivated by the strangest symbols of happiness.

But it isn't just the taste of the water that matters. It is also the sound!

How can you navigate when they crowd the waterfields with titanic shadows and nerve-shattering echoes? And they have so many names for them: *Freighters* and *Tankers* and grand *Ocean Liners*? The size of storms that darken the above!

Do you believe how they worship their giants? Those they call *Carriers* and *Great Ships of War*, peace-loving *Destroyers* waltzing through waters they claim for themselves.

What would you say if you heard them boasting their clever technologies designed to lay waste? Discarded parachutes, torpedo control wires; transducers and cables that crisscross ocean floors; eavesdropping sonobuoys that sink down and spy?

To you, their insidious madsounds rival the rest of their weaponry: steel undershadows that skulk through the bigwater, piercing your eardrums with shrill pinging screams.

Won't the two-leggeds consider how difficult it is for your kind to communicate when background noise masks what you hear and forces you to call louder to be heard? Your senses don't include echolocation like that of the toothed whales, but is it a surprise that intense sonic waves rained down from the shadows might have grave effects on your sensitive physiology?

Who can hear anything in such a ruckus?

Is that the laughter of their machines?

Why do their sleek black undershadows pretend to move like cetaceans before screaming *PINGGG!* with a vengeance? Do they really believe in their right to bend water with shockwaves that drive so many kinds of whales into madness? Or don't they give a thought to what their exploding decibels reap beneath the watersky? Are they trying to batter your cells and rupture your blood vessels? Or is it just a matter of bad luck?

Do the two-leggeds suspect *why* whole families of marine mammals become so desperate to escape the torture that they launch themselves out of the depths into the realm of dry death? The toothed whales most of all; *Odontocetes*, the two-leggeds call them—the dolphins and beaked whales and sperm whales and orcas. Blasted with sonar, inner compasses shattered, leaving them spinning in mad figure eights . . . until they have suffocated by waterdeath or cast themselves ashore.

The two-legged "*why?*" makes no sense to you. That is not your word. It belongs to them. One of their sounds for the seeking of answers.

Their redtops use it all the time. *Why?* Because they have had to chronicle the strandings of so many mammals of the waterfields.

Like intellectual vultures they have picked over your rotting carcasses on the sand to study your hemorrhaged eyes and mutilated ears. They have recorded the blood clots throughout your hulks, the trauma to your brains, the inflammations of your tissues, the lesions in your spoiled kidneys, the congestion smothering your lungs.

In some whales they have analyzed the symptoms of decompression sickness, the nitrogen bubbles that can wreck the blood and bend a body in two.

And still they ask, *why?*

The two-leggeds call themselves "man," as if a title allows them to rejoice at your massacre. Don't they realize that they are the victims, that they are part of nature too?

You would never demand answers, but can they not see their laughter for what it really is? *Man'slaughter?*

As you swim through the swaths of waterfields they call *shipping lanes*, and beyond the great cape they refer to as Hatteras, do they neglect to question, *How DO you swim? How DO you . . . DO?*

How do you move the length of an island with one slow sweep of your gargantuan tail?

How do you glide with such power while conveying an energy they call humility?

To think that two-leggeds may miss the moment to see you throw your broad flukes, lifting your dark form with the brawn of an earthquake! Have they forgotten the thrill of watching you explode through the watersky into the realm of the wings of the air? Are they wasting the chance to hold their breaths as you fly suspended in clouds to view past, future, present—all at a glance?

Have they closed their eyes to the awe of your dance?

Don't they wonder, even if it only lasts for a second, how does one feel to be touched by light?

CHAPTER SEVEN
THE CURIOUS MOUSE

Along thinwater coasts of dune, scrub, and palm that two-leggeds call Georgia Bight, you recognize the brushing sound of aquamarine wavelets on sandy shores.

You glimpse familiar curtains of sunlight undulating across the corals and islands that beckon you in your own language, scattered like gardens across the brightwater.

At last, you have returned to the birthwater, guided by your inner eye.

Your brothers and sisters have also arrived, weary and hungry, in largo movements of their bittersweet sonata. Yet, they vocalize an urgent sadness throughout the endangered sanctuary of their calving grounds.

Several are hobbled by knotted ropes, frayed netting, and plastic buoys. One trails a shadow's length of groundline from his jaw, now a mess of shredded baleen and torn flesh.

The gory sight doesn't surprise you. Three quarters of your kind have been entangled at some point, their lesions and gashes and damaged bones all testament to their ordeals.

Others are missing. And there is no sign of the pregnant mothers.

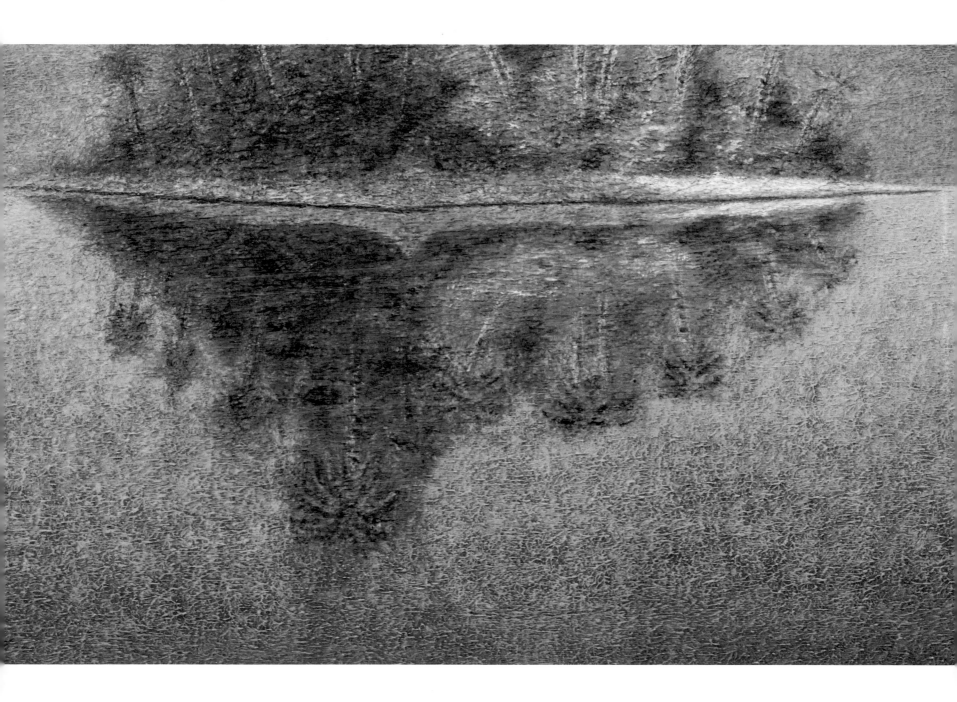

Do you wonder what frights they might have encountered that would prevent them from reaching the bliss of the birthwater?

It has always been a magical place for your kind. Is that why your fasting begins and ends here?

When the Mother Light drifts down to sleep, you thrust up to the watersky and savor the waves lapping along the great length of your back. You roll to one side and watch the pastel clouds fade, awaiting the twinkle of silver sparks in the satin blue Between.

After several shallow dives, an unknowable force stirs you to return to the depths.

Why do you always salute the above when preparing to vanish for twenty minutes on your freefall into the lightless realm? Each time, you take a deep breath and arch your great back with purpose in your momentum. Then your flukes rise and hover like the wings of a butterfly, before you slip away.

Maybe it really does give you a chance to transform yourself, dreaming of another life yet the same as this.

When the Mother Light rises, you spoon the surface of the watersky, opening your great dark eye as wide as an obsidian crystal ball.

You look for the two-legged redtops on the beach, as if they might be worried about your missing sisters. But you have no understanding of whether two-leggeds care that you hang by a thread far weaker than the byssus tethering a mollusk to rock. Have you ever heard them protest that they should not be deprived of their freedom to take what is rightfully theirs?

Often you notice the dryshadows zooming past, shiny as barracudas and spouting blue fumes. They leave tracks in the sand as proof of their dominance. Loud trundling creatures with stubby round flippers, their crowns are as clear as sea jellies.

Do you ever hypothesize about those organs you see inside?

Are they the brains or hearts? Or simply no more than the guts?

Certainly you are not the only one that matters, you might reason—if whales do such a thing. After all, every creature has to make a living. Everyone has to eat.

But how can you eat? How do you live through year after year of two-legged clatter from fishing and shipping, sonar, construction, exploring for black gold in every lost ocean? In the course of each longwater have you ever guessed *why* pods of you have been haunted by stress? How are you supposed to find food in that racket?

For the Between's sake, how are you supposed to find . . . *each other* ?

How can your grandmothers pass the word about where to forage? Where to dive down? Where changing salinity marks fresh water plumes? Where touch or temperature or sounds give you clues? In so much noise, can your calls be heard to remember topography, landmarks of the deep? Where zooplankton concentrate in thick enough patches to make it worth your energy to feed?

You may not echolocate prey the way other whales or dolphins do, but to catch squid or albacore you don't have the speed. And, for their fat or protein, you have no need.

Do you wonder if two-leggeds have a clue about nutrition or what you really require to eat well?

So what if you are one of the largest animals on the planet, though you dine on creatures the size of fleas?

Does it strike you that two-leggeds seem to hunger for more? Do you notice that some *shadows that move with thunder* do so in the name of defending their shores.

> *Their shores? They kill to take all they want from shores they declare their own, until nothing is left for the taking. Where is the sense in that?*

You'd never ask something so foolish, but in all the stories your grandfathers passed down, did they ever mention earthfields or waterfields that *belonged* to any creature? Would your grandmothers have believed the two-legged system called *fisheries*, and the ghost shadows that have dragged the waterfields bare?

Or does the miserly belief in ownership explain why two-leggeds have plucked away so many named with the two-legged sounds? The *cod* and the *hake*; the *whiting* and *pollock*; the *halibut*, *herring*, *haddock*, and *mackerel*; the *grouper* and *snapper*; the *redfish* and *skate*; the *swordfish* and *ray*; the *shark* and the *marlin*; and even the lightning-fast *tuna?*

Does it make you conclude the two-leggeds have forgotten that every creature speaks and has something important to offer? Is that why the list builds like angry seas before demonic winds, cloaking the world below, including the millions of your brethren, the great mammals of the bigwater: beluga and bowhead; blue and gray; fin and minke; sei and humpback?

Or do the fates of those creatures reflect the two-leggeds' strength and purpose, not just what they wish to eat? Who are you to question two-legged judgment when they measure intelligence by rules they have written?

No doubt, the *shadows that move with thunder* are capable of showing how strong they really are, powerful enough to fight back against the most threatening foes.

You can't conceive how horrible those must be.

If you thought about it, would you ponder what could possibly be the most dangerous creature the earth has ever known?

You may not be bold enough to persuade the two leggeds to change. But then, your meekness is nothing new. That is why they call you one of the *Mysticetes* —you are the whale they have baptized "the mouse."

Chapter Eight
The Above

Now, in the warmth of the birthwater, you feel compelled to call out for your youngest daughter. Her unexpected reply eases your worries for a moment. But the tone of her voice sounds amiss. Though she has grown beyond the time of nursing, she cannot sleep because of the nightmarish fishing gear wrapped around her jaw, the frayed lines trailing her flukes like tentacles. Maybe that is why she has sought you, to be reassured by swimming together again.

A flash of movement rivets your senses. Suddenly, a finned predator weaves a bold course toward your daughter, interested in anything wounded or weak that wanders alone when the Mother Light sleeps.

You warn it off with a bellow as your daughter sings out in fear.

The enormous shark makes a second run, grimacing to reveal rows of pearly stilettos.

You brush it away with a mere flick of your tail.

Bruised and unwilling to waste precious energy, it sulks back to the depths in search of easier prey.

Does it curse you as sharks do? Perhaps. Still, you eye it coolly until it has vanished.

Does the painful sight of your daughter disturb you to remember her healthy and unharmed, dappled in clearwater light? Or do the tremors in the turquoise waves murmur that you must not let down your guard? Since reaching the birthwater, you know the *shadows that move with thunder* have been gathering.

You have heard hosts of them, clouding the watersky. Steaming relentlessly across the bigwater, not just in pursuit of *umbrellas* anymore, but all manner of strange sounds for what the two-leggeds desire: *iron* and *rubber; pulpwood* and *coffee; computers, televisions, clothing,* and *sporting goods; semi-conductors, footwear, tobacco; crude oil* and *soft drinks; aircrafts* and *poultry; carpets* and *toys* and *nuclear fuel; writing* and *art* and *medical supplies; automobiles* and *jewelry* and *wine* . . .

Storming the currents with noise, the madsounds have come again.

They begin like slow waves against a shore, faint as the distant cry of a gull.

They keep building, until unnerving pulses roll toward you like bloodcurdling moonlight.

What does that look like to a whale? Oh, if only the two-leggeds knew.

The pulses grab your attention. First, you get goosebumps, followed by shivers, building in magnitude to chills, then cold fright—from two-legged blasting, dredging, drilling, and intense seismic bursts that pummel the seafloor.

But mostly it comes from the merciless shadows, flooding the waterfields with *noise*.

Sudden as a peel of thunder, an invisible force batters you and your daughter. The watersky darkens. Your daughter lurches sideways from shockwaves, choking in snarls of groundlines and gear.

A bullwhip cracks through her head, almost loud enough to perforate her eardrums. The *noise*.

It takes no time at all. It takes all of forever. The *noise!*

Her brain throbs and blisters, her petrified heart pounds. The tangles of gillnet and lines choke tighter.

Scarlet waves flood her mind with her grandmothers' screams. It is their horror she feels as she wails like a banshee, split in half by the pain, the deafening noise shredding her to pieces.

You urge her to flee, but the nooses squeeze taut. The sizzling pitches are cooking her mind. She is too young for this! She can no longer detect your warning call sounds.

NMMOOOUUU! she screeches, her flesh curling back.

NMMOOOUUU! she weeps as if bleeding inside.

You leap and dive to put yourself between the source of the sonic torture and your daughter. It is no more a dance of wonder, but a writhing last lunge.

The *shadows* crowd the watersky like schools of greatfish. The din becomes a blunt bone that hammers your lungs with *noise*.

Lightning bolts singe your brain as the sounds explode through you.

Dragon claws blare; the *noise* gouges your flanks. You feel as though it is pealing your blubber, ripping it off.

You thrash your tail, arch your back.

The *noise* turns to lava, evil magma, scorching your gullet to cinders.

You struggle in the death-grip of an iron vise when your entire body catches fire in *noise*.

It is no use. Do you say that to yourself?

No use?

But your daughter is shrieking! *Does no one else hear?* The shadows bear down overhead. She moans for pity, begging for the end, to silence the agony.

DOES NO ONE ELSE HEAR!

She whirls and plunges to escape the waterbending, to shake free from the cutting lines that tighten more with every movement. Until she is overtaken by desperation.

Sobbing frantically, she rockets upward as you call to her.

NMMOOOUUU! you cry to stop her flight—but you already know. *Too late, too late.*

With an earth-shaking thrust of her wide black flukes she launches into the above, shattering the watersky as if she might reach the Mother Light after all.

Then she is *flying*! Up there! Can you believe your eyes?

You roll over to watch her soaring through clouds as free as a dream.

Until the dream splinters to a vision of hell.

She crashes back through the watersky where the same demons await, sapping her strength until she can barely move. The demons bent on her utter undoing.

She drifts through the thinwater on her last ounce of breath, rumbling in pain, wheezing for mercy, frozen by terror, her sobs bubbling away into nothingness.

Her strained final breath sounds of chains being dragged through crushed gravel and seashells, as the waves gently nudge her to the dry realm's edge.

In silence.

Do you know how the redtops found your daughter on the beach today? Driven mad by entanglement, burned up by noise. Entrapped in razor rope, her flesh gouged, her bones sawed through. In the growing red twilight her limp flukes curled skyward. But she was already gone.

Her life fluids spilled out into rivulets that snaked and vanished in the burning sharp sand. Her internal organs were bursting under their own burden. Her feverish core was passing the boiling point.

The steam from her hulk rose like ghosts toward the first spark of a bright yellow star.

You raised your head to look on from the shallows, mourning for a child caught up in needless destruction where once lived innocence and beauty. But all you could make out was her charred silhouette languishing in fire and blood.

Does it seem like a *war*? Of course, that's not one of your words. It doesn't occur to you that this happens each day in the two-legged world, where they have grown to depend on their wars.

Can the two-leggeds know how you actually feel? Trapped between traffic and nooses and noise?

No. They cannot. But they could take a guess. They could pause for a moment, ask, *What does this mean?* That would be something. If only, if only . . .

In your weakest moments of rage, or caught by a painful flash of your daughter, your black eye blazes at the two-leggeds.

Do you beseech them in anguish?

YOU! You forget me! You forget all of us!

Do you question their capacity for denial?

How many creatures have you treated this way? With all their wisdom to offer, how many have you banished

from the circle?

Does your bleak mind spin toward them in resonant sorrow or in challenge?

WILL YOUR GRANDCHILDREN CURSE YOU FOR EMPTYING THE WATERFIELDS

OF MINE?

It appears that only a few redtops are there to listen.

How can you disregard the two-leggeds' arrogance that poisons the womb of their own mother? Can you pity them if they presume this is also their right? Or is that just an anthropomorphic reflection, easier for two-leggeds to bestow on you than on sharks or sea urchins or cephalopods?

The answer doesn't matter to you. It is only the perception of continuity for which you feel the urgent need, the chance that future grandmothers will sense your presence in their memories.

Are you capable of forgiveness, Arpeggio? If you experience anger, will it ever fade?

Those questions don't matter to you, either. Though perhaps, in a sense, you do forgive them. For cynicism is not a part of your soul.

Yours is a being of trust—no less than the stars on blue-black nights. The stars shine with faith that their twinkling will never cease; come dawn, it is completely absorbed into the supernal glow of the Mother Light.

CHAPTER NINE
THE FIRST WALTZ

Some of the two-leggeds do think your voice has beauty, you know. They say it carries the balance within you and throughout the oceans, like the balance that used to be . . .

of birth and death,

earth and sky,

water and light.

A few believe that you still dream of a dance with your daughter, that you sing with the joy and sorrow of your grandmothers. Even though all of them are gone.

Your grandmothers knew it as the song of the waterfields. A sound that is no less than whale and water, or the dying of broken oceans. No more than love or hate, hope or despair. No greater than the first waltz that two-leggeds once sang and danced with all the earth.

Do the redtops ask what their sciences are worth, if two-leggeds cannot find their humanity with which to apply the knowledge they have gained? What good are their clever ideas if they cannot learn how to dance the dance again?

Will they ever learn to ask, *When IS an ocean not just an ocean?*

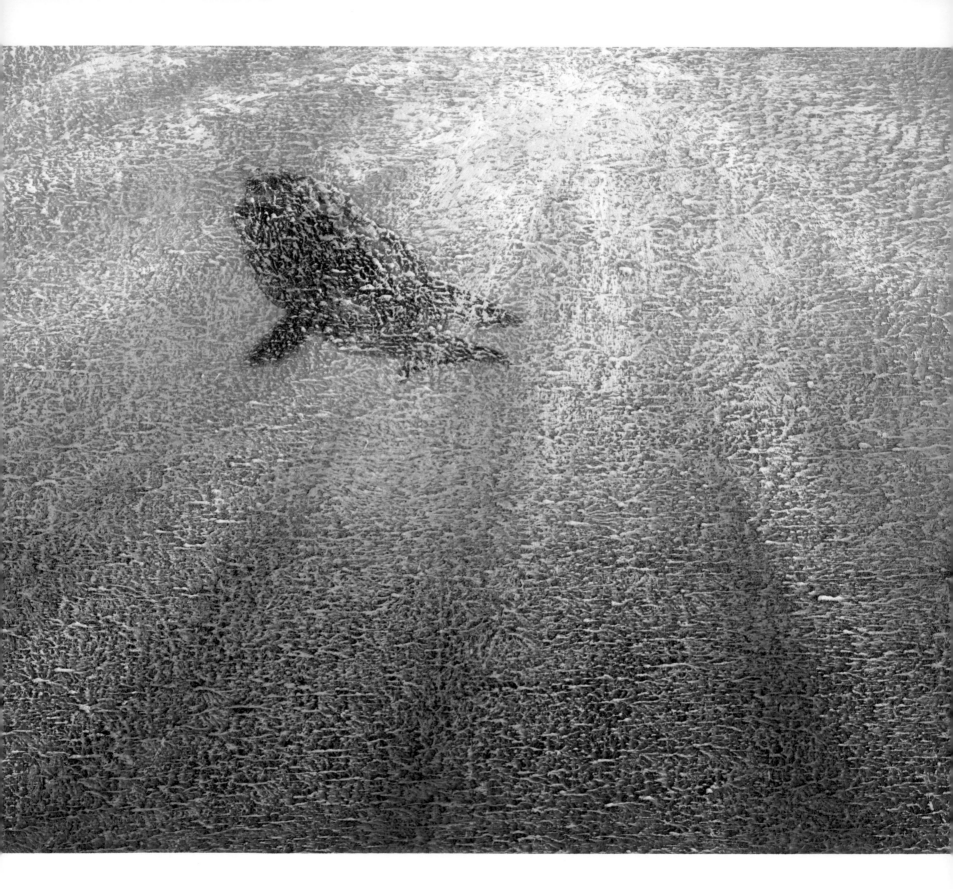

When it is a cloud or a raindrop, that feeds a forest and fills a river, flowing on and on, to become part of a moss or a wetland, a plankton, a coral, a mussel or lobster, a kelp or a herring, a gull, an osprey, a salmon, a marlin, a squid or a tuna, a bear or a seal, a penguin, a krill, a whale?

Or simply the womb of a living blue planet, gasping for one last breath?

What use is the sheer miracle of your presence? Do two-leggeds even consider this wisdom? Do they ask what role a whale might play in the watersky's living web?

Do you wish you could toss back an irreverent answer?

> *Me? My role? What about you? What good are YOU? Have you thought about that?*
> *WHAT GOOD ARE YOU!*

Your senses have no time to respond, interrupted by a childlike wail from far out in the birthwater . . .

> *What's this? What's this!*

It hits you like the breaking of a wave—the quicksilver flash of a minnow in a tide pool!

Allegro and dolce, fast and cheerful and sweet as ocean dew, the blissful sound stops you where you float.

Motionless, you drift . . .

You listen with every spark of your soul . . .

You wait . . .

Until a timid flame flickers in your consciousness.

Gradually, a bright surge of lightness fills you up like waves of nectar from the moon.

Would the two-leggeds call it gratitude, if they were capable of this feeling?

The question doesn't occur to you.

You veer away from the dry realm as though turning your back on a shadow. Moving with effort—for it does take strength to carry the leaden weight of sorrow in your memory—you head deeper into the birthwater.

In this moment, you are overcome with that thankful wave of lightness. The young ones have made it! You hear them again, their beautiful sonorous lowing! They have arrived! The longwater has brought them to safety one more time.

You roll as you reply, slapping your flukes on the watersky with a thunderclap that echoes throughout the waterfields. How many are these new young voices? Beyond the masking, you cannot tell.

All you know is a ripple of fulfillment as pure as moonbeams, coursing through your great body—the perception of . . . enduring.

Maybe this alone is your answer, your deepest desire; that a new calf will be the next who waltzes with giants in the blue waterfields.

Indeed, Arpeggio, imagine if two-leggeds did stumble across such a glorious secret. In the twilight of oceans—however they perceive it, before dawn or after dusk—wouldn't this be a wise beginning, changing their dream of how the world should be?

Do you think the children of your children might have another shot at existence after all?

THE FIRST WALTZ   87

Imagine!

Then wouldn't you have something to sing about!

AFTERWORD
THE RIGHT WHALE RESEARCH TEAM, NEW ENGLAND AQUARIUM

What does the future hold for the North Atlantic right whale?

Since the early 1980s, research and monitoring programs in known habitat areas have provided valuable data to assess right whale life history and human impacts on population trends over time. The information has been used by researchers, government managers, conservationists, and members of the fishing and shipping industries in the United States and Canada to develop, implement, monitor, and assess management strategies. The North Atlantic Right Whale Consortium is a unique example of the cooperative work of these groups to advance recovery efforts for this highly endangered species.

Presently, North Atlantic right whales have more protection than ever from the impacts of human activities on their survival. Shipping lanes have been changed and areas to be avoided have been implemented in eastern Canada. Along the eastern seaboard of the United States, in addition to similar changes in shipping lanes and areas to be avoided, vessel traffic has been slowed in high risk areas. These measures were put in place to reduce the risk of vessel strike mortalities. The fishing industry in the United States has been required to reduce the amount of rope in the water column by shifting from floating to sinking groundline in addition to a suite of other measures to reduce the risk of entanglement. Voluntary efforts in Canada are underway as well.

The impact of ocean noise from shipping, seismic, and construction activities continues to be a major concern for this species as it can affect their ability to communicate and could lead to chronic stress. Efforts to both understand the effects of ocean noise on these animals and to find ways to reduce ocean noise are in progress but regulatory measures have yet to be developed.

During the next decade, the right whales will show us if mitigation measures enacted thus far have been effective at reducing human-related mortalities from shipping and fishing. Although still critically endangered, the trend in population growth suggests the species could recover if the whales continue to reproduce at the current rate and mortality is adequately countered with the mitigation measures. We can proudly propose this scenario because of the hundreds of scientists, government officials, conservationists, and concerned citizens, internationally, who have dedicated themselves to changing how humans use the oceans to give this species, at risk of extinction a few decades ago, a second chance.

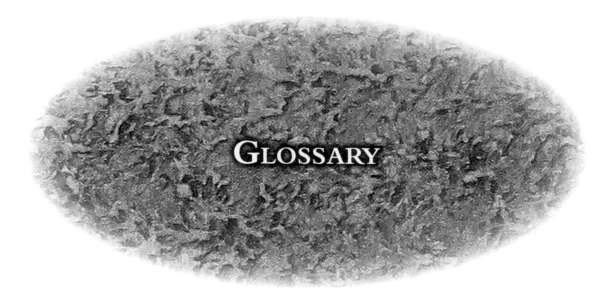

GLOSSARY

above: the realms beyond the interface of watersky

adagio: quite slow

albatross: among the largest wings of the air (family Diomedeidae), wide-ranging in the Southern Ocean and North Pacific; once present in the North Atlantic

allegro: fast and cheerful

anemone: or sea anemone (order Actiniaria), a polyp that resembles a flower, often attached to rocks or sand underwater

anthropomorphic: described as having human characteristics

Arpeggio: the name given to a female North Atlantic right whale born in 1997, New England Aquarium North Atlantic Right Whale Catalog #2753; last sighted in 2011 (as of this printing)

arpeggio: broken chord in which the individual tones are sounded one after another instead of simultaneously

balance: sustainable equilibrium

baleen: strands of keratin that hang from the rostrum of a baleen whale and strain plankton from water

basso profundo: deep bass singing voice

Bay of Fundy: body of water between Nova Scotia and New Bruswick and the feeding ground where the greatest numbers of *E. glacialis* right whales have been observed in recent decades

Being: existence; life

bends, the: decompression sickness, caused by excess nitrogen or other gases in the bloodstream

Between: the infinite space between the stars; also, the transcendent state of existence; also, the place where a soul rests before choosing a body for its next life

bigwater: the ocean

birthwater: calving grounds

blackwater: the depths beyond the reach of light

blubber: the thick layer of fat between skin and muscle of marine mammals

bonnet: the usually large callosity found on the front part of the head (or rostrum) of a right whale

boarding knife: a knife with sharp edges on both sides, like a sword

breedwater: mating grounds

brightwater: tropical water of light hues

bycatch: fish, turtles, marine mammals, or seabirds unintentionally caught while fishing for other species

byssus: the threadlike filament that certain mollusks use to attach themselves to hard surfaces

callosities: thickened skin patches, often cream-colored due to infestations of whale lice; found above the eyes, near the blowholes, on the jaw (or mandible) and chin, and along the rostrum

calving grounds: area in the oceans where marine mammals give birth

carnivore: an animal that eats flesh

caudal: near the tail; posterior

cetacean: marine mammal, of the order Cetacea, that includes dolphins, porpoises, and whales

circle, the: the round of seasons; temporal and eternal aspects of life

clearwater: any part of the oceans with extensive visibility

contralto: the lowest of the female voices

copepod: zooplankton; the primary prey of the North Atlantic right whale is the calanoid copepod, *Calanus finmarchicus*

cyamids: crustaceans that populate the surface of callosities; whale lice

deadwater: an area of the ocean devoid of pelagic life forms

deathline: any lobster pot rope, polysteel line, or long-line used for lobstering or fishing that hangs or drifts in the water column

decibel: a unit of measurement of intensity for sound or signal

diminuendo: growing softer

dissonance: combination of tones that sound discordant and unstable, in need of resolution

dolce: cheerful and sweet

dorsal: relating to or near the back

Dream Light: the moon

dreamworld: the dimension of experience beyond that of the five senses; transcendent reality of vision, trance, dream or spirit

dry death: suffocation by stranding

dry realm: land above the high water mark

dryshadows: vehicles on land

duet: musical piece for two voices or instruments

earthfields: the lands, and their associated energies; the dry realm

echolocation: biological sonar

effluent: something flowing out; a sewage discharge

eight-legged crustacean: American lobster (*Homarus americanus*), also known as *northern lobster*, *Atlantic lobster* or *Maine lobster* (the lobster actually has ten "legs," although the first two are called claws and not used for movement)

entanglement: the entwining or wrapping of fishing gear, lines, nets, or ropes with a whale or other marine creature

Eubalaena glacialis: North Atlantic right whale; in Latin, the name means "good" or "true" whale of the ice

farwater: a distant destination

field: an area of activity or influence

flensing: the stripping away of blubber or skin

flotsam and jetsam: any objects found floating in the waterfields or washed up in thinwater or the dry realm

fluke: either half of the tail of a whale

foodwater: area with dense enough concentrations of prey for a whale to feed

gestation: pregnancy

gillnet: a fishing net that hangs in the water and entangles fish by the gills

groundline: the rope between baited fishing or lobstering traps that rest on the bottom

greatfish: whale or other large deepwater creature such as marlin or pelagic shark

habitat: an environment in which a species or ecological community normally occurs

hemorrhage: to bleed a lot

hippopotamus (*Hippopotamus amphibious*): a semi-aquatic land mammal of sub-Saharan Africa

interlude: a period of time; a short musical segment between parts of a longer piece of music

keel: principal structure of a ship that runs its full length along the centerline

kegging: the practice of using floating barrels attached to a kiltusk to keep a whale on the surface

killing tide: harmful algal bloom (HAB) with biotoxins, sometimes called "red tide"

kiltusk: harpoon or iron (made of a barbed iron tip and shaft mounted on a wooden pole); also called a flue iron or "harping iron"

krill: shrimp-like crustacean that is prey for many species of whales

landshadow: motorized vehicle found in the dry realm

largo: in a broad and dignified manner; very slow

legato: smooth and connected; opposite of staccato

libretto: the text of an opera or other dramatic musical composition

logging: sleeping

Long Cold: the most recent glacial period or Ice Age

longline: commercial fishing line with baited hooks fastened by short branching lines at various distances

longwater: the annual migration; full turn of the seasons; chronological reference to an approximate year by the two-legged calendar

loon: aquatic bird (*Gavia*) like a duck or a goose

madsounds: any sound waves (often high decibel) that block out natural levels of sound, inducing panic, pain, or causing physical damage, depending on the species being affected

man'slaughter: manslaughter

masking: the obscuration of one sound of interest by other interfering sounds

minuet: slow, stately dance

Mother: the Earth

Mother Light: the sun

mystical: having a spiritual reality or meaning not apparent to the intelligence or senses of two-leggeds

Mysticetes: cetaceans in the suborder Mysticeti, the baleen whales; "the mouse"

New-Found-Land: Newfoundland; Canada's easternmost province, in the region of the right whale's historical feeding grounds

nitric: including or coming from nitrogen

nodding: behavior of cleaning off baleen after feeding

Odontocetes: cetaceans in the suborder Odontoceti, the toothed whales

plankton: small organisms that float in large numbers, often at or near the watersky

pectoral: relating to the chest or breast

pod: a school of marine mammals

polypropylene: a plastic polymer; often used for food containers, some kinds are extremely strong and durable

ram: as in "ram filter feeding," the manner in which balaenids (right and bowhead whales) swim forward and open their mouths to feed

realm: a territory or kingdom; a field of influence

realm of dry death: a marine mammal perception of any land above the high water mark

redtop: scientist, often a whale biologist; so called because of a bright orange or red foul-weather top or personal flotation device (PFD)

requiem: a musical composition or ceremony for the dead

rostrum: upper jaw

serpent: two-legged buoy line or groundline in the water column

SAG: surface active group

Scotian: of or relating to Nova Scotia

shadow: vessel on the surface of the watersky

shadows that move without thunder: wind-powered vessels

shadows that move with thunder: engine-powered vessels

ship-strike: collision between a vessel and a whale

smallshadow: skiff or small boat; commonly, a motorized inflatable

sonata: instrumental genre in several movements for soloist or small ensemble

sonobuoy: disposable sonar system used for underwater acoustic research or anti-submarine warfare

sparks: stars

staccato: characterized by a series of detached sounds or elements

stranding: beaching of marine mammals or sea turtles in shallows or on shore

thinwater: coastal waters or shallows

thundershadows: engine-powered vessels

trawl: a fishing net dragged behind a thundershadow; two or more lobster pots strung together

two-legged: a human being

undershadow: submarine

violoncello: bowed-string instrument with a middle-to-low range and dark, rich sonority; lower than a viola

water column: two-legged conceptual column of water from the surface to the bottom

waterbender: a creature that uses sonar to navigate or locate predators or prey

waterdeath: drowning

waterfields: the oceans and their associated energies

watersky: the marine mammal perception of the surface of the ocean; surface of any body of water

wise one: an elder creature or two-legged that possesses wisdom; a two-legged scientist

zooplankton: plankton that consists of animals

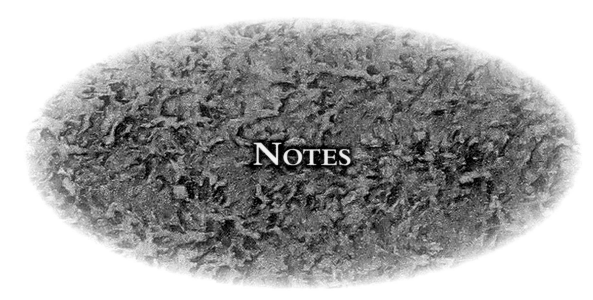

NOTES

Foreword

Page x **They cover more than 70 percent of the earth and contain nearly all her water . . .**

See National Oceanic and Atmospheric Administration. *Ocean.* Washington, D.C.: www.noaa.gov/ocean.html.

Page x **And in the 3.8 billion years of life on this planet . . .**

See Center for Biodiversity and Conservation, American Museum of Natural History. "Biodiversity in the Next Millenium Survey." New York: *American Museum of Natural History*, 1998: http://cbc.amnh.org/crisis/crisis.html. See also American Museum of Natural History Press Release. "NATIONAL SURVEY REVEALS BIODIVERSITY CRISIS–SCIENTIFIC EXPERTS BELIEVE WE ARE IN MIDST OF FASTEST MASS EXTINCTION IN EARTH'S HISTORY." April 20, 1998: www.well.com/~davidu/amnh.html.

Page x **By the 1990s, . . . an estimated 27,000 species were dying off each year.**

See E. O. Wilson. *The Diversity of Life.* Cambridge: Belknap Press, Harvard University, 1992: p. 280.

Page x **Twenty percent of earth's species were predicted to be lost by 2030 . . .**

See "NATIONAL SURVEY REVEALS BIODIVERSITY CRISIS," op. cit.

Page xi **Near the end of the millennium it was calculated that more than 137 species . . .**

See www.rain-tree.com/facts.htm.

Page xi **. . . *half of all species* may be extinct by the century's end . . .**

See E. O. Wilson. *The Future of Life.* New York: Knopf, 2002: p. xxiii.

Page xi **Considering the undeniable interconnectedness of life forms throughout the six kingdoms . . .**

By the late sixties, biologists described five organized kingdoms (proposed by Whittaker) or taxonomic classifications: Animalia, Plantae, Fungi, Protista, and Prokaryota or Monera. (David Kirk, principal author, Cecie Starr, editor. *Biology Today, Second Edition.* New York: CRM/Random House, 1972, 1975: p. 37). That system is still found in many textbooks in Britain, Australia, and Latin America, whereas textbooks in the United States now use a system of six kingdoms: Animalia, Plantae, Fungi, Protista, Archaea, and Bacteria. As of 2010, "there is no set of kingdoms sufficiently supported by current research to gain widespread acceptance" (http://en.wikipedia.org/wiki/Kingdom_(biology)).

Page xi **"When we destroy ecosystems and extinguish species . . . "**

See *The Future of Life*, op. cit.: p. 39.

Page xi **And why does the work of scientists and environmentalists sometimes appear . . .**

See Neal Evernden. *The Natural Alien, Humankind and Environment, Second Edition.* Toronto, Buffalo, London: University of Toronto Press, 1985, 1993, reprinted 1999: pp. 8–11.

Page xii **. . . since the 1950s, ninety percent of the large ocean fish are gone.**

See Ransom A. Myers and Boris Worm. "Rapid worldwide depletion of predatory fish communities." *Nature* 423, 280–283. London: Macmillan Publishers Ltd., May 15, 2003: www.nature.com/nature/journal/v423/n6937/full/nature01610.html.

Page xii **. . . by 2010, over three hundred thousand dolphins, porpoises, and whales were dying as bycatch . . .**

See www.worldwildlife.org/what/globalmarkets/fishing/bycatch.html.

Page xii **" . . . The acquisition of knowledge always involves the revelation of ignorance . . ."**

See Wendell Berry. *Standing by Words.* Washington, D.C.: Shoemaker & Howard, 2005: p. 65.

Page xii **Only recently have we begun to unravel the engineering marvels of mussel adhesives . . .**

See Janine M. Benyus. *Biomimcry: Innovation Inspired by Nature.* New York: Harper Perennial, 2002: pp. 118–136.

Page xii **The burgeoning field of biomimetics looks to the wisdom of living systems . . .**

See Ibid., p. 59.

Page xii **" . . . it does not matter whether it can be scientifically proved that life as we know it . . . "**

See Carol Jacobs, Cayuga Bear Clan Mother. *Presentation to the United Nations July 18, 1995.* www.ratical.org/many_worlds/6Nations/PresentToUN.html. Reproduced from *Akwesasne Notes New Series,* Fall, October/November/December, 1995, Volume 1, #3 & 4: pp. 116–117.

Page xiii **" . . . we remain largely ignorant of how the infinitely complex connections . . . "**

See Paul Hawken. *The Ecology of Commerce.* New York: HarperBusiness, HarperCollins 1993, paperback 1994: p. 27.

Page xiii **. . . the North Atlantic right whale (*Eubalaena glacialis*) . . .**

For good overviews of facts about this animal, see: www.acsonline.org/factpack/RightWhale.htm; www.learner.org/jnorth/search/RightWhale_notes1.html; or
   www.neaq.org/animals_and_exhibits/animals/northern_right_whale/index.php.

Page xiii **It likely earned its name because baleen . . .**

See Randall R. Reeves and Robert D. Kenney. "Baleen Whales: Right Whales and Allies." *Wild Mammals of North America: Biology, Management and Conservation,* edited by George A. Feldhamer, Bruce C. Thompson, and Joseph A. Chapman. Baltimore and London: The Johns Hopkins University Press, 2003: p. 425; cited in Michael P. Dyer, "Why black whales are called right whales." Forthcoming publication, 2012. Common usage has attributed the name to the idea that, because of its valuable oil and baleen, and the ease with which it was approached and killed, this was the "right" whale to hunt. That etymology has been reexamined, and there are several theories about how the misusage of the name "right whale" evolved.

Page xiii **Where many thousands of North Atlantic right whales inhabited the seas . . .**

See B. A. Mcleod, M. W. Brown, M. J. Moore, et al. "Bowhead Whales, and Not Right Whales, Were the Primary Target of 16th- to 17th-century Basque Whalers in the Western North Atlantic." *Arctic,* Vol. 61, No. 1, Mar 2008: pp. 61–75. North Atlantic right whale populations were once thought to be in the tens of thousands prior to commercial whaling. Recent research suggests this number is lower. It has been estimated that 5,500 to "possibly twice that number" of North Atlantic right whales were killed "between 1634 and 1950." See R. R. Reeves, T. D. Smith, and E. Josephson. "Near annihilation of a species: Right whaling in the North Atlantic." In: Scott D. Kraus and R. M. Rolland, editors. *The Urban Whale: North Atlantic Right Whales at the Crossroads.* Cambridge, MA: Harvard University Press, 2007: pp. 68–69.

Page xiv **But other factors have also contributed to their decline . . .**

See Toolika Rastogi, Moira W. Brown, Brenna A. McLeod, et al. "Genetic Analysis of 16th-Century Whale Bones Prompts a Revision of the Impact of Basque Whaling on Right and Bowhead Whales in the

Western North Atlantic." *Canadian Journal of Zoology*, (2004), Vol. 82, Issue 10. NRC Research Press: pp. 1,647–1,654.

Page xiv **Environmentalists often appear to be working only reactively to undo the impacts . . .**
See Reynolds III, John E. "The Paradox of Marine Mammal Science and Conservation." *Marine Mammal Research, Conservation beyond Crisis*, Baltimore: The Johns Hopkins University Press, 2005: p. 2.

Page xv **" . . . new technologies and new habits offer some promise . . . "**
See Bill McKibben. "Carbon's New Math." *National Geographic*. Washington, D.C., October 2007: p. 36.

Page xv **"There's no such thing as sustainable technology or economic development . . . "**
See Alan Weisman. *Gaviotas: A Village to Reinvent the World*. White River Junction, VT: Chelsea Green, 1998: p. 13.

Chapter One Imagine

Page 2 *Shhhhhh . . . Listen!*
Look again. In the painting, *Moonwaves*, you may see a silhouette beneath the waves on the right side of the canvas.

Chapter Three Remembrance

Page 16 **You shimmer and glow like a candle brought to life . . .**
Reference to a mystical experience of deep meditation; *The Bhagavad Gita*, ch. VI, v. 19: "When meditation is mastered, the mind is unwavering like the flame of a lamp in a windless place."

Page 17 **. . . encoded by genes passed on to you by your ancestors . . .**
See Richard Dawkins. *The Greatest Show on Earth, the Evidence for Evolution*. New York: Free Press, 2009: p. 170.

Page 23 **They came from the eastern shores of the bigwater . . .**
A reference to the early Basque whalers who crossed the North Atlantic seeking more plentiful populations of whales. They and their galleons eventually made their summer home in places such as Red Bay, Labrador, which became, in the sixteenth century, the world's whaling capital.

Page 23 **Their boarding knives and spades and flensing blades . . .**

For a good overview of whaling tools, see: www.whalecraft.net.

Page 24 **"*Black whale!*" they called out.**

A common name (among whalers) for the three species of *Eubalaena*: North Atlantic, Southern, and Pacific. "Why black whales are called right whales," op. cit.

Page 24 **. . . and the waterfields began to rise and consume the dry realm.**

See NASA/Goddard Space Flight Center. "International team to drill beneath massive Antarctic ice shelf." *ScienceDaily*. November 9, 2011. Retrieved December 8, 2011, from www.sciencedaily.com/releases/2011/11/111109194323.htm. For background on Sea Level Rise and Climate change, see: www.climate.org/topics/sea-level/index.html.

Page 27 **Perhaps you have communicated to others that the warmongering thundershadows . . .**

Whether the impacts of two-legged activity can be registered throughout a species is one of many questions put to the reader. Consider the true story of an individual (cataloged later as Eg #1045) that endured wounds from high-powered rifles, a ship's propeller, probable infection, and constant pain. "If Eg #1045's experiences with humans became known among her kind," suggests Scott D. Kraus in *The Urban Whale*, "right whales would surely avoid us all." *The Urban Whale*, op. cit.: p. *3*.

Chapter Four The Undoing

Page 34 **You startled at the jab, and tried to move away.**

For a detailed description of this kind of scenario, see: Moore M, Walsh M, Bailey J, Brunson D, Gulland F, et al. "Sedation at Sea of Entangled North Atlantic Right Whales (*Eubalaena glacialis*) to Enhance Disentanglement." PLoS ONE 5(3), 2010: e9597. doi:10.1371/journal.pone.0009597; http://www.plosone.org/article/info:doi%2F10.1371%2Fjournal.pone.0009597.

Chapter Five Waterfields

Page 42 **. . . until some of them lost their way in the darkness of their own eyes.**

Reference to a beautiful phrase by Oglala Sioux Holy Man, Black Elk: "It is in the darkness of their eyes that men get lost." See John G. Neihardt. *Black Elk Speaks, Being the Life Story of a Holy Man of the Oglala Sioux*. Lincoln, Nebraska: University of Nebraska Press, 1961: p. 2.

Chapter Six Man's Laughter

Page 46 A fish once so plentiful that its schools could stop the shadows in their tracks!

Reference to the cod fishery in Newfoundland and Labrador from around 1550 to 1610. Explorer Sebastian Cabot wrote famously about the cod that swam in schools so bountiful that " . . . they sumtymes stayed his shippes." See William Broaddus Cridlin, Secretary of the Virginia Historical Pageant Association. *A History of Colonial Virginia: The First Permanent Colony in America*. Richmond, Virginia: Williams Printing Company, 1922: Chapter 2; cited in *New River Notes*, Historical and Genealogical Resources for the Upper New River Valley of North Carolina and Virginia: www.newrivernotes.com/va/cridlin1.htm, accessed March 3, 2010.

Page 48 . . . especially with the distinctive vee shape of your plume.

Reference to a sacred perspective as much as to a sign of the presence of a right whale. For many indigenous cultures, an open V is also a symbol of the power of spirit to use its intelligence creatively and positively. See Heike Owusu. *Symbols of Native America*. New York: Sterling, 1997: p.17.

Chapter Seven The Curious Mouse

Page 62 . . . another life yet the same as this.

See Peter C. Stone. *The Untouchable Tree, An Illustrated Guide to Earthly Wisdom & Arboreal Delights*. New York: Skyhorse Publishing, 2008: p. 112.

Page 69 . . . you are the whale they have baptized "the mouse."

Reference to the origin of the word Mysticetes, the baleen whales. "New Latin, *mysticētus*, from Greek *mustikētos*, alteration of (*hō*) *mus to kētos*, (the) whale (called) the mouse." See *The American Heritage Dictionary of the English Language*, Third Edition. Boston and New York: Houghton Mifflin Co., copyright 1996, 1992: p. 1,196. Describing baleen, Aristotle (384–322 BCE) wrote, "The so-called mousewhale instead of teeth has hairs in its mouth resembling pigs' bristles." See the following: *The Works of Aristotle* translated into English Under the Editorship of J.A. Smith M.A. and W.D. Ross M.A.*, Volume IV; *HISTORIA ANIMALIUM* by D'arcy Wentworth Thompson. Oxford: Clarendon Press, 1910: Book III, Chapter 12; http://etext.virginia.edu/toc/modeng/public/AriHian.html. A variation in the etymology apparently derives from sailors noting the resemblance of baleen to a mustache. Thus, the suborder Mysticetes becomes "*Mystacoceti* (from Greek μυσταξ 'moustache' + κητος 'whale')." See James C. Mead and Robert L. Brownell, Jr.'s "Order Cetacea," November 16, 2005, pp. 723-743, in Don E. Wilson and DeeAnn M. Reeder, eds. *Mammal Species of the World: A Taxonomic and Geographic Reference, 3rd Edition*. Baltimore: Johns Hopkins University Press, 2 vols: p. 2,142; cited in http://en.wikipedia.org/wiki/Baleen_whale#cite_note-1.

Chapter Eight The Above

Page 71 **. . . what the two-leggeds desire:** *iron* **and** *rubber,* *pulpwood* **and** *coffee* **. . .**

See www.worldsrichestcountries.com/top_us_imports.html.

Page 72 **From two-legged blasting, dredging, drilling . . .**

See Susan E. Parks and Christopher W. Clark. "Acoustic Communication: Social Sounds and the Potential Impacts of Noise." *The Urban Whale,* op. cit.: p. 326.

SELECTED BIBLIOGRAPHY

Clapham, Phil. *Right Whales, Natural History and Conservation*. Stillwater, MN: World Life Library, Voyageur Press, 2004.

Johnson, T. *Entanglements: The Intertwined Fates of Whales and Fishermen*. Baltimore, MD: University Press, 2005.

Kraus, Scott D. and Kenneth Mallory. *Disappearing Giants, The North Atlantic Right Whale*. Charlestown, MA: Bunker Hill Publishing in association with the New England Aquarium, 2003.

Kraus, Scott D., and K. Mallory. *The Search for the Right Whale. A New England Aquarium Book*. New York: Crown Publishers, 1993.

Kraus, Scott D., and R. M. Rolland, eds. *The Urban Whale: North Atlantic Right Whales at the Crossroads*. Cambridge, MA: Harvard University Press, 2007.

Mann, Kenneth, and John Lazier. *Dynamics of Marine Ecosystems: Biological-Physical Interactions in the Oceans*. Hoboken, NJ: Wiley-Blackwell, 2005.

Payne, Roger. *Among Whales*. New York: Charles Scribner's Sons, 1995.

Reynolds, J. E., and S. A. Rommel, eds. *Biology of Marine Mammals*. Washington, DC: Smithsonian Books, 1999.

Reynolds, J. E., et al. *Marine Mammal Research; Conservation beyond Crisis*. Baltimore: John Hopkins University Press, 2005.

Twiss, J. R., Jr., and R. R. Reeves, eds. *Conservation and Management of Marine Mammals*. Washington, DC: Smithsonian Books, 1999.

Selected Resource Links

American Cetacean Society: www.acsonline.org/factpack/RightWhale.htm

Journey North, Right Whale: www.learner.org/jnorth/search/RightWhale_notes1.html

New England Aquarium: www.neaq.org/animals_and_exhibits/animals/northern_right_whale/index.php

NOAA Fisheries Northeast Regional Office: www.nero.noaa.gov/

North Atlantic Right Whale Catalog: rwcatalog.neaq.org

Northeast U.S. Right Whale Sighting Advisory System (SAS): www.nefsc.noaa.gov/psb/surveys

Partners of the North Atlantic Right Whale Consortium: www.narwc.org/index.php?mc=1&p=8

Rightwhale.ca: www.rightwhale.ca

Right Whale Listening Network: www.listenforwhales.org

RightWhaleWeb: www.rightwhaleweb.org

Whale and Dolphin Conservation Society International: www.wdcs.org

WhaleNet: whale.wheelock.edu

Pen & Inks

Chapter 1.   *North Atlantic Right Whale (Eubalaena glacialis)*

Chapter 2.   *Blue Whale (Balaenoptera musculus)*

Chapter 3.   *Narwhal (Monodon monoceros)*

Chapter 4.   *Beluga Whale (Delphinapterus leucas)*

Chapter 5.   *Bottlenose Dolphin (Tursiops truncatus)*

Chapter 6.   *Humpback Whale (Megaptera novaeangliae)*

Chapter 7.   *Sperm Whale (Physeter macrocephalus)*

Chapter 8.   *Killer Whale, Orca (Orcinus orca)*

Chapter 9.   *Calf, North Atlantic Right Whale (Eubalaena glacialis)*

ACKNOWLEDGMENTS

I am immensely grateful to the many people who contributed at various points to this project:

Marcella Matthaei, who first asked me to tell the story of the North Atlantic right whale;

Michael Moore at Woods Hole Oceanographic Institution, who steered me in the right direction;

The scientists and research assistants with the New England Aquarium who graciously offered feedback and answered countless questions: Moira Brown, Cynthia Browning, Jonathan Cunha, Marianna Hagbloom, Kathleen Hunt, Philip Hamilton, Amy Knowlton, Scott Kraus, Kerry Lagueux, Marilyn Marx, Kate McClellan, Tracy Montgomery, Heather Pettis, Rosalind Rolland, Jessica Taylor, Amanda Thompson, Jodie Treloar, Tim Werner, and Monica Zani;

The many other scientists I've met at North Atlantic Right Whale Consortium Conferences;

Bill McWeeney (reputed by his students to be a contemporary of Darwin) and the Calvineers at Adams School, Castine, Maine;

Lauri Murison, and Sarah and Allan Macdonald, aboard the *Elsie Menota*, Grand Manaan Island, New Brunswick, Canada;

Michael Dyer, Maritime Curator, and Bob Rocha, Science Director, of the New Bedford Whaling Museum, a partner for this book with its educational programming goals in areas of history, language arts, culture, and science;

Tess Cederholm, Virginia Freyermuth, Bud Ris, and David Welty, who kindly read incomplete versions of the manuscript;

The devoted educators with whom I've worked on innovative curriculum design in recent years, including Jack Crowley, Virginia Freyermuth, Tracy Higgins, Laurie Robertson-Lorant, Hugh O'Mara, Mary Ellen Nochimow, Jessica Ross, David Welty, and our students, from whom we never stop learning;

My editors, Julie Matysik and Bill Wolfsthal, for their keen eyes and guidance;

And, my son, Oliver, who accompanied me on several adventures by land and by sea, following the migrations of the North Atlantic right whale.

Finally, I offer thanks to my father, who saw some of the paintings while the book was underway, although he never had a chance to read it. His love of oceans, boating, fishing, snorkeling, and scuba diving, and his great advocacy of research and education, no doubt charted a course for this story through another lifetime.

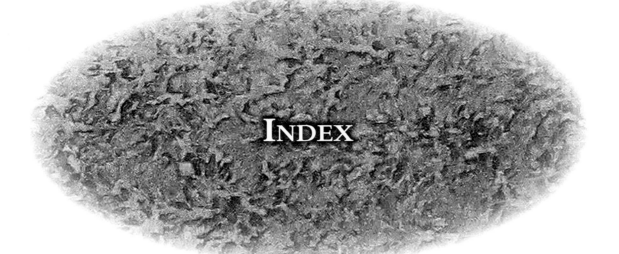

INDEX